Y0-BVO-132

New Modes of Local Political Organizing:

Local Government Fragmentation in Scandinavia

New Modes of Local Political Organizing:

Local Government Fragmentation in Scandinavia

Peter Bogason (Editor)

Nova Sciences Publishers, Inc.
Commack

Art Director: Maria Ester Hawrys
Assistant Director: Elenor Kallberg
Graphics: Denise Dieterich, Kerri Pfister,
Erika Cassatti and Barbara Minerd
Manuscript Coordinator: Sharyn Schweidel
Book Production: Tammy Sauter, and Benjamin Fung
Circulation: Irene Kwartiroff and Annette Hellinger

Library of Congress Cataloging-in-Publication Data

New modes of local political organizing: local government fragmentation in Scandinavia/ Peter Bogason (ed.).
p. cm.
Includes bibliographical references and index.
ISBN 1-56072-266-5 :
1. Local government--Scandinavia. 2. Decentralization in government--Scandinavia. I. Bogason, Peter.
JS6142.N49 1996 95-48993
352.048--dc20 CIP

6080 Jericho Turnpike, Suite 207
Commack, New York 11725
Tele. 516-499-3103 Fax 516-499-3146
E Mail Novasci1@aol.com

Printed in the United States of America

Contents

Foreword vii

1. **Changes in Local Government and Governance**
   Peter Bogason 1

2. **The Modernization of Local Government in the Modern Democratic State**
   L J Sharpe 15

3. **The Political Division of Labor: An Institutionalist Approach to Authority and Community**
   Henrik Paul Bang 35

4. **Local Institutional Change in Sweden - A Case Study**
   Stig Montin and Gunnar Persson 71

5. **Decentralization, Privatization and Representativeness in Local Government**
   Henrih Bäck 93

6. **Community Councils - Improved Local Democracy?**
   Allan Dreyer Hansen, Torill Nyseth & Nils Aarsæther 115

7. **Learning User Influence. The Constitution of User-Subjects in a Differentiated Welfare State**
   Hans Wadskjær 143

8. **The Fragmented Locality**
   Peter Bogason 169

Index 191

Foreword

Peter Bogason

This book on government changes in Scandinavia has emerged from a conference on Local Institutional Change at Roskilde University in February 1994. The conference was sponsored by the Center for Local Institutional Research, a temporary center for networking scholars with an interest in institutional analysis. The center is partially funded by the Danish Social Science Research Council.

A number of presenters were asked to revise their paper for publication, and one chapter was commissioned from a non-participant to strengthen the theoretical discussion on politics in the community.

This book should be of interest to students and researchers in local government and institutional analysis, mainly within political science, public administration, urban affairs and sociology. The empirical focus is on Scandinavia, but the theoretical implications go far beyond those particular countries. The development towards fragmentation of local governmental bodies seems to be universal, at least in the Western countries. This calls for reconsideration of our conceptual luggage which has mainly been linked to formal organizations; the development seems to challenge such an understanding in favor of broader concepts like governance, to replace government as a basic unit of analysis. Formal government is not obsolete, but does not suffice any longer.

Roskilde University
January 1996
Peter Bogason

Chapter 1

CHANGES IN LOCAL GOVERNMENT AND GOVERNANCE

Peter Bogason

This book is about changes in governance at the local level in Scandinavia - Denmark, Norway and Sweden. The concept of "governance" rather than "local government" is used to stress the theoretical focus which is less on formal organization and more on emerging patterns of interaction. Even though much of what is reported is related to changes in formal organization, the actual consequences are challenging to a traditional notion of local government. The formal local governments are counties and communes and subdivisions of communes like neighborhood councils. Often the generic term commune is used for all sub-national governments, and the term municipality is used for the primary level. Thus, there might be some variation in the use of the term in this volume - but the context should make the interpretation clear to the reader.

This introductory chapter has two purposes. First, it gives the reader an overview of the local government reforms that have taken place over decades in the three countries. This is meant as background information only; the references will take the reader further, if desired. Second the chapters of the book are summarized as a reader's guide to his or her special interests.

1. Recent Local Government Reforms

For several decades, most local governments in North Western Europe have experienced a growth in tasks and resources. In the Scandinavian countries, this has been because local governments have administered most of the functions of the welfare state: social security, social services, health and schools; functions that have grown in expenditures because of a more fine-tuned coverage system. Of total 1992 public consumption, local governments spent 69 per cent in Denmark, 60 per cent in Norway and 70 per cent in Sweden (Nordic Yearbook of Statistics 1994: table 198). Table 1 details some of the major public functions.

Table 1. Local government share of major public consumption functions.

	Denmark	Norway	Sweden
Education	63%	76%	79%
Health	93%	93%	99%
Social security & Welfare	95%	83%	92%
Culture & recreation	60%	66%	90%
Transport & communication	66%	46%	46%
Housing & community amenity	50%	N/A	82%
Public order	9%	16%	18%

Source: Nordic Yearbook of Statistics 1994, table 198.

It is extremely difficult to give a valid, comparative, statistical measure of local government expenditures (Bogason 1987). There are many ways of counting expenditures: are they total expenditures, expenditures out of town sources, or what? There is no general agreement on such matters. Therefore, other sources may give somewhat different details. Still, the general pattern indicated here seems to hold true no matter what source is used; at least the ranking of the countries. Local governments dominate the functions of education, health, social expenditures, and culture. They play a small role in public order and safety, defense and industrial development (not shown in table). In housing, they play a role in Sweden, in Denmark and in Norway, communes have a surplus from public housing incomes.

As big spenders in the localities, local governments are also big employers. Schools and hospitals are staff-intensive, and since many local services like care for the elderly and for children are run by local governments, they may be the largest employer in some areas. Full-time two-income families time are the rule rather than the exception in Scandinavia, and hence there is a high demand for such services.

Below we shall discuss some of the most important changes as an empirical-analytical basis of understanding for the chapters that follow.

1.1. Scandinavian Reforms

In the 1950s, 1960s and 1970s, the national governments in Denmark, Norway and Sweden carried out rather sweeping local government reforms to put them in a position to administer those tasks efficiently. Consequently, local governments were merged into larger entities; they had their tax-base improved simply by getting more population and by primarily basing it on income taxes rather than e.g. property taxes. The state grant systems were tuned to supplement the local tax-base and to equalize differences between rich and poor areas of the countries.

The three countries, however, differ somewhat[1]. Sweden which had reforms in the 1950s and again in the 1970s, has the largest communes averaging about 30,000 inhabitants. Norway has the smallest with an average of about 8,000 and Denmark falls between, averaging 18,000. The Danish and Swedish communes have stronger local income taxing powers than the Norwegians. The organizational settings differ somewhat, but in general hospitals and health care are regional tasks, most day-to-day services including primary schools are commune responsibilities. At the regional level, the Swedes have a stronger state presence and hence more mixed state/county council projects and administration than do the Danes and the Norwegians. They were more determined to sort out tasks as either state or county responsibility at the regional level at the time of reform.

The reforms of the sub-national government were probably seen as less dramatic by the city inhabitants than in the rural population. If one were to take a typical Scandinavian rural community in the 1940s, it was a sort of parish commune size with 500-1500 inhabitants where many people knew one another personally and the local elected parish commune council took care of the few destitute in the area. The council had no professional staff but a part-time secretary; schooling beyond the elementary

[1] For details of the reforms and the general status of local government in the three countries around the year 1990, see Bogason 1990, Gustafsson 1990 and Hansen 1990.

level was taken care of by inter-communal arrangements with the nearest town etc. Those wanting a better education had to move or enroll in a boarding school. In other words, of the rural communities needs for public services were perceived as few and life was fairly simple, as far as public responsibilities were concerned.

The mergers changed all that. The working operations of sub-national governments completely changed. It went from organizations mostly reacting *ad hoc* to personal demands when presented by the individual in person, into sophisticated policy-making systems. Communes were expected to set up policies for their activities; planning became the accepted view of solutions to all problems, at least in the political/administrative rhetoric. Local administration was centralized to the center town of the new commune; in the rural areas new bureaucracies were nearly built from scratch; in the towns the existing bureaucracies expanded. At the same time, national legislation was fine-tuned to cover more and more aspects of daily life. As a result, the rural areas got opportunities for public assistance and services no one had dreamed of in the 1940s. And in the cities, the supply of public services also grew rapidly[2].

The need for mechanisms to coordinate local activities grew, and step by step, the managerial and policy-making staff changed from individuals trained as office clerks to business and university graduates. Street-level bureaucrats increasingly had professional training. And the role of the councillors changed along the way, from local lay individuals involved in case-work, to party politicians deciding complex operating plans in committees of the new large communes. Those representatives became less and less involved in those concrete matters which were in the mind of the voter who called his or her local councillor for information on work-in-progress.

In the mid-1980s criticisms began to be voiced. The commune administrative bureaucracies had grown too large. One side of the criticism followed a public choice line of thinking, maintaining that the bureaucrats were too responsive to alleged citizen demands; that they converted the perceived demands into demands for more professional staff and therefore grew beyond political control. This criticism led to demands for better political control with increasing levels of expenditure and followed

---

[2] This is not to say that demands grew solely because of the new organization systems, but rather than without the new organizations, the commune systems would have collapsed and consequently state agencies probably would have taken over. In the rural areas, however, the social changes we have seen were undoubtedly helped by the capabilities of the new communes.

several lines of advice to improve managerial skills and make politicians interested in setting goals and priorities for levels of activities within policy fields. Another side of the criticism followed a line of democratic theory holding the view that the bureaucracies really were unresponsive to the true demands of the citizens who consequently were getting alienated to the commune system of governance. Here the remedy was to get more citizens involved in decision-making, either by setting up (again) more neighborhood councils or by constructing new channels for user influence.

In the Scandinavian countries the criticisms, of course, followed somewhat different lines according to the national systems of commune government. Let us first take a look at the problems of political control.

In Denmark the general pattern of organization was one of a Commune Council as the general forum for the elected representatives; the council then typically had standing committees to run the actual administrative affairs within sectors like social affairs, schools, roads and transport, and urban zoning code. Sweden and Norway also had a general commune council plus statutory administrative committees, but in addition they had quite a large number of fairly independent boards appointed by the council, but seated by non-councillors. They took care of local policy fields like recreation, culture and libraries. Mostly the boards members were members of the political parties, though. Whatever the concrete form, all three Scandinavian systems encouraged a specialization among councillors and led to a compartmentalized administrative structure with competition for resources rather than coordination between the service functions as a result. Some observers saw the commune council composed of the school party, the social party etc. rather than by Social Democrats, Conservatives and other political parties.

The answers to these problems, of course, differed very much in their concrete forms, and they certainly have not been solved by all communes. Those which have actually attacked the problems by organizational measures, all intended to strengthen coordinating mechanisms. They can be characterized in a general way as follows.

One solution was to abandon most of the specialized committees and merge the administrative sections into larger entities. This made the administrative layer just below the top more lean so that budgetary competitions among chiefs were reduced. This was the Danish way which was faithful to the basic principles of the reforms, but made some adjustments which made administrative incentives work better for an integration (Andersen 1992, Larsen 1993). The Norwegians took two steps (Baldersheim 1993:159- 60): one is quite dramatic by introducing the principles of the national cabinet into the municipal council: naming "ministers" responsible for administration and subject to a possible vote of no confidence and hence removal from office. The other - more popular - model is a business-like strategy of delegating powers to chief administrative officers and making them accountable to the commune council as managing directors to the shareholders of a company.

All three strategies in their own way stressed the political accountability as something somehow involved in running daily affairs. The Swedes, however, went much further, by making it possible to sever the organizational ties between politics and administration completely. They introduced a provider-producer distinction; the political council has the role of the provider setting up goals for the service, and any public (or, for that matter, private) organization can have the role of producer fulfilling the goals by having a contract with the provider, and specifying conditions for service production.

So much for the organizational pattern promoting general coordination and overview for the politicians. But as we saw above, there also were criticisms regarding the increasingly passive role of the citizen-at-large in community matters of the amalgamated communes. The organizational reaction was somewhat slow (Kolam 1987:9-11). Norway came first; Oslo got neighborhood councils in 1973, and other areas followed. After a period of reluctance and outright resistance from the advisers for central government, some communes in Sweden met critics by setting up advisory neighborhood councils in 1969, local administrations in 1974 and neighborhood councils with decision-making powers in 1980. Denmark came last and only recently amended its legislation to formally allow communes to establish such councils with real powers, if they so wish[3]. The strongest contestant has been the city of Copenhagen, and therefore the largest city in Denmark was exempted until 1996 from the possibility of setting up neighborhood councils! There are differences among the countries in rules for appointing members; in Sweden and Denmark they mirror the political composition of the commune council. Sweden has not implemented a clause of direct local council elections. Thus there may be tensions between the majority of the neighborhood council and the community population in highly segmented communes, e.g. with distinct areas with high-rise buildings as opposed to garden city areas for the well-off.

[3] Thus the Danish examples of Neighborhood councils referred to in the chapter below by Hansen, Nyseth and Aarsæther are based on specific state approvals.

Denmark has introduced another channel for citizen influence: the organizational user board. Public schools and child day care centers are required to set up Board of Directors where representatives elected by the parents are in majority. The boards are responsible for the running affairs of the organization, with certain limitations that may vary commune by commune. They cannot, however, influence the freedom of teaching nor the basic, state regulated curriculum. Such boards probably will proliferate in other fields; homes and day care centers for the elderly are now encouraged to introduce them.

Neighborhood councils and user boards both put stress on local influence, but neighborhood councils have a coordinating function among local activities while user boards only reflect the needs of users to control the particular activities within the organization. Thus completely different participatory mechanisms are at work; the former encouraging the generally interested citizen to work for the community, the latter helping the user improve a specific service.

How come so many ideas have been implemented in local governments within a rather short period of time? One explanation may be found in the rather large-scale free commune experiments which took place in the second half of the 1980s. The idea was conceived by the Swedes, and Denmark and Norway quickly followed. Briefly stated, the free commune experiments allowed local governments to disregard state regulations - by special state permission! - and thus initiate experiments to enhance administrative efficiency and/or citizen influence (see e.g. Baldersheim & Ståhlberg (eds) 1994). So they were allowed to lax administrative procedures, to disregard certain requirements, to establish new boards, to merge statutory committees, to coordinate separate computerized data bases, to delegate authority from committee to professionals, and to avoid approval of loans from the state prefect, etc.

Whether or not these experiments were successful is not the issue here; the important thing is that taken together, the experiments gave a basis for changing the way many things were done in the communes. The experimental period can be regarded as a sort of collective learning process by trial and error. Not only communes, but also state agencies learned lessons that led to changes in legislation and/or administrative law.

1.2. Trends in Local Government

Several factors of change in local Scandinavian government are quite similar - they concern organizational efficiency; new conceptions of the people using services; and finally users get new roles in managing service organizations. Details, however, do differ.

First, the general management of local governments is increasingly oriented towards managerial efficiency of limited means. The central governments try to restrain either the spending propensity or the revenue possibilities of local councils which therefore must perform their services more efficiently. The differences between having responsibility for provision of local services and actually producing them are stressed by different versions of contracting out; in Sweden, the provider-producer scheme. The two other countries do contract out, but there is no central policy on the issue. In addition, local public service organizations get more responsibility for conducting their affairs, including having greater budgetary autonomy.

Second, the role of the direct user of public services is slowly changed to include the "customer" as well as the "citizen". People are increasingly involved in services as purchasers, not only as local residents with a need. At least, this is how the rhetoric evolves; we have yet to see people being screened according to purchasing powers. But there is a clear tendency towards individualizing the relations between service producer and user within several types of service.

Third, in some services a number of users are put into a managerial role to serve - on behalf of the whole group of users - on a board of directors or trustees. These elected users are supposed to go beyond their individual interests and run the service to the benefit of as many users as possible. But their role is restricted to that particular service; they have no responsibility whatsoever to the community as such. But the Scandinavians keep the formal integrating role of the commune council, the tendency of e.g. the British to take more and more of these roles away from local general purpose councils (Stewart & Stoker (eds) 1979) is not a Scandinavian feature.

Thus the changes should not be interpreted as elements of a general undermining of local government as service provider in the community. In the Scandinavian countries, there might have been - particularly in the 1980s -some suspicions of a sinister central power wanting to cut back the local governments, but in general, the changes have taken place in relative harmony between the center and the periphery. In Scandinavia, there is a continued faith in local politics, but under renewed circumstances.

2. CHAPTER OUTLINES

The chapters in this volume address the changes outlined above in different ways, giving the reader a wide array of responses to the general question of what is happening and what are the consequences. Of course, there is no single and simple answer. The way things develop differs among countries and even among localities within countries. Within the limits of this volume, the reader is offered representative evidence of change, but there is no attempt to cover the full ground of alterations in all three countries. That would require one volume per country.

Sharpe's contribution (chapter 2) sets the larger parameters for the development at the local level. As a general pattern, local government has been enlarged in Western Europe, particularly in the North. In the North, municipalities are service producers and hence subject to economies of scale. In the South there is no obvious link between communes and a particular range of services. Hence there are no reasons for reorganization in the South due to economies of scale. But the claims that economies of scale are obtained are in themselves dubious, based, as they are, on the theory of the firm, and even there put forward as an axiom rather than as an empirically tested truth. Anyway, it has not had profound impact on the size of many municipalities, even in the non-Napoleonic countries. Consequently, local government size is a political question for a political entity.

Sharpe makes it clear that participation is a questionable parameter for local government size, since the tightest bonds would be between 50 people or less. In addition, extremely small communities may suffer from oligarchic and discriminating tendencies. If we switch the focus of interest from population size to geographic size or distance, however, there is a clear link in terms of accessibility of services. And finally, a point against large sub-national entities is that a desirable sense of local identity may be lost if the boundaries of the local governments spans many different communities of interest.

Henrik Bang analyzes the development of political activity in the locality (chapter 3). Using a survey of trade union members as the empirical basis, Bang calls for theoretical analysis based on the praxis of citizens by translating new forms of cognition developed in the political interaction between leaders and lay-actors into new theoretical discourses. These are based on the accomplishments of citizens who are able to transform the parameters of daily life in the community rather than just subscribe to solutions demanded and managed by the political elite.

Regarding analyses of the locality, Bang questions the need for taking departure in the concept of the state; new forms of choice are being formed by people in the communities enabling some of them to act with new forms of autonomy. Bang goes along with the new institutionalism and combines it with Easton's understanding of the political system. He challenges the conceptual suitability of the instruments of control related to models of the state, the market and the civil society. The challenge for democracy is not to make the state more efficient or responsive, but to bring political leaders and citizens back into governance in a world where more and more resources are controlled by non-governmental elites. This is done by taking part in community affairs so that every citizen, however marginalized, must be thought of as bearing some responsibility for why things are the way they are locally.

The two Swedish chapters illustrate problems related to the introduction of sub-communal political bodies. After two waves of amalgamation in the 1950s and 1970s, Swedish politicians realized that there was a need for more political community-level organization and set up a framework legislation to permit for such bodies. Furthermore, contracting out and the introduction of user vouchers became fashionable as a way of establishing a quasi-market.

Montin and Persson's chapter on local institutional change in Sweden (chapter 4) highlights problems between the local "center" political bodies - commune councils and the political parties - and neighborhood democracy. A first step in the municipality of Örebro was to set up neighborhood councils - but even those are of a considerable size, covering about 10,000 inhabitants, and they mirror the political seats in the municipality council, not the area they serve. A number of user boards have been set up to advise the neighborhood council. Their members are partly elected by people in the localities, and partly appointed bureaucrats from the neighborhood administration. It appears that many people in the locality - particularly non-public employees and business people - still see the boards as dominated by "system preferences" rather than genuine local ideas.

Still, an alternative - a quasi-marketization of public services - has not really been successful in those areas. Montin & Persson maintain that people in Örebro prefer local - but not necessarily party political - decisions by elected boards rather than determination by consumer preferences.

In the chapter on decentralization, privatization and representativeness in local government (chapter 5), Bäck analyzes the consequences of primarily two kinds of changes in local politics - the introduction of neighborhood councils and the contracting out of service production - for democratic representation in terms of the number and social composition of councillors. It seems clear that in areas where neighborhood councils have been established, the number of politicians has increased, and vice versa in areas where contracting out has been a major feature of modernization. And since there are fewer politicians in those areas, multiple duties are more frequent there compared with neighborhood council areas . Finally, areas with "market solutions" are doing relatively poor in terms of deprived group representation .

In other words, neighborhood councils seem to increase the possibilities for contacts between citizens and politicians in the localities, while contracting out and other quasi-market solutions give fewer possibilities for such contacts.

Hansen, Nyseth and Aarsæther compare four community councils in Norway and Denmark with respect to their role in improving local democracy (chapter 6). They see four distinct possibilities for improving local democracy by such councils: act as a catalyst and create an institutionalized local public sphere of debate; make possible citizen participation in community matter decisions; channel local opinions into the agenda both in the neighborhood council and higher levels of elected councils; and broaden participation indirectly functioning as a recruitment basis for higher levels in the political system. Community councils may, however, also have some dysfunctions like only promoting the local elite and its opinion, promoting parochialism and reducing the role of party politics.

The four cases show that community councils promote local participation by allowing for more contact with citizens in regular meetings; by recruiting new types of citizens and by institutionalizing more open citizens' meetings. Thus the councils brings in a more active citizenry, particularly in the rural areas, and there is not a similar loss at the commune council level: this is not a zero-sum game.

Hans Wadskjær discusses some of the problems related to increased self- governance in services for the elderly, illustrated by a case study, and primary schools (chapter 7). Both policy fields are strongly influenced by professionals and semi-professionals, acting as a mediating level between the political level and the users. For decades they have been responsible for developing and fine-tuning basic welfare state services. They have set up quite sophisticated channels of influence by means of professional interest organizations that have dealt with the national leadership and the interest organizations of local governments. They have created and preserved autonomy for the street-level bureaucrats in a non-adversarial way; professionals were seen as creating the true and only solutions to welfare state problems.

Wadskjær demonstrates how these mediators tend to react rather negatively when user governance is introduced. At the local level, street-level bureaucrats have seen user influence as an attack on the autonomy created over the years. The immediate reaction has been frustration and as little active involvement as possible on the part of the professionals and semi-professionals. The case study, however, indicates that change takes time, and it seems possible over time to reach an understanding that is of mutual benefit to users and staff. To reach this understanding, an open dialogue is seen as necessary; and this is a radical break away from the rather bureaucratic traditional way most professionals think. The political level was a necessary component in the process to positively sanction the development; but the users also had to show some muscle. A similar understanding has not yet been accomplished in many public elementary schools which, one should add, are more complex organizations than a service center for the elderly which forms the backbone of the empirical research in this chapter.

The editor rounds off the volume by discussing the fragmentation of governance in the locality (chapter 8). A tendency towards fragmentation is found in all the countries. They have come about in an experimental mood involving both national and local governments in an non-adversarial way. The public fragmentation is paralleled by fragmentation in the production and family spheres, the hallmarks of the post-modern society with much compartmentalization and little centralizing power. The individual is in charge, but free to engage in collective processes; there is potential for the politicization of most aspects of life, but also to live in peace and quiet. The fragmentation is a challenge for the traditional management views since the organization is opened up for a complex interorganizational process where traditional means of organizational control are of little help. It is also a challenge for the citizen in so far as more active participation will be required for those who want to have influence on the public services. On the other hand, this is precisely the antidote to the public welfare bureaucracy which has developed, requiring little involvement by the clients or anybody else, but therefore also having the potential for alienation.

REFERENCES

Andersen, Helle Sundgaard. 1992. "Ændringerne i de politiske og de administrative roller i det danske kommunalpolitiske system." *Nordisk Administrativt Tidsskrift* 73(4):328-40.

Baldersheim, Harald. 1993. "Kommunal organisering: Motar sel, men ressoursar afgjer?" In *Organisering av offentlig sektor. Perspektiver - reformer - erfaringer - utfordringer*, ed. Per Lægreid and Johan P. Olsen, 155-68. Oslo: TANO.

Baldersheim, Harald, and Krister Ståhlberg, eds. 1994. *Towards the Self-regulating Municipality. Free Communes and Administrative Modernization in Scandinavia.* Aldershot: Dartmouth.

Bogason, Peter. 1987. "Capacity for Welfare: Local Governments in Scandinavia and the United States." *Scandinavian Studies* 59(2):184-202.

------. 1990. "Danish Local Government: Towards an Effective and Efficient Welfare State." In *Local Government and Urban Affairs in International Perspective. Analysis of Twenty Western Industrialised Countries*, ed. Jens Joachim Hesse, 261-90. Baden-Baden: Nomos.

Gustafsson, Gunnel. 1990. "Swedish Local Government: Reconsidering Rationality and Consensus." *In Local Government and Urban Affairs in International Perspective. Analyses of Twenty Western Industrialised Countries*, ed. Jens Joachim Hesse, 241-60. Baden-Baden: Nomos.

Hansen, Tore. 1990. "Norwegian Local Government: Stabiliity Through Change." In *Local Government and Urban Affairs in International Perspective. Analyses of Twenty Western Industrialised Countries,* ed. Jens Joachim Hesse, 211-40. Baden-Baden: Nomos.

Kolam, Kerstin. 1987. *Lokala organ i Norden 1968-1986. Från idé till verklighet.* Umeå: Statsvetenskapliga Institutionen.

Larsen, Henrik. 1993. "Nye Organisationsformer I Kommunerne." *Nordisk Administrativt Tidsskrift* 74(2):7-30.

Stewart, John, and Gerry Stoker, eds. 1989. *The Future of Local Government.* London: MacMillan.

References

[illegible]

Chapter 2

The Modernization of Local Government in the Modern Democratic State

L J Sharpe
Nuffield College, Oxford

Introduction

This paper* deals with some fundamental aspects of decentral government in the modern democratic state, ie., what is the case for modernizing its local government system. The paper sets out the causes of such modernization including the key factors of suburbanization and the service revolution. Both tend to lead to the enlargement (population and area) of local government units. But not everywhere; some states have clearly been more assiduous modernizers than others. One reason for this disparity is the difference between, on the one hand, the Southern European states with their napoleonic system of central-local relations which for various reasons makes local government modernization politically very difficult. On the other hand, in the North European states modernization of the local government system was very much easier to undertake. Another

* This paper is drawn from *Local Government: Size and Efficiency and Popular Participation*, (Strasbourg: Council of Europe, 1994)

factor influencing the extent of local government modernization is the correctness of the enlargement thesis. In short, does increased scale provide the predicted functional benefits? This question is closely explored together with the related question as to whether scale is beneficial for popular participation. The question is posed: is the reverse true? That is to say, is small scale democratically beneficial because democracy is a diminishing function of scale. Finally, an additional possibility is raised which places a large question mark against the enlargement solution. This is the link between small scale and popular access to local services. It will be seen that the paper has thus to plough some already well-tilled fields and apologies are offered in advance for such repetition. However, despite the fact that we are dealing with a fundamental set of institutions and concepts, nevertheless such things are not immune from fashion and so there is always a need for a re-statement of, even, the obvious.

I. The Modernization of Local Government

The central concern of this paper is the complex question of altering local government structure - which usually means enlargement - to meet the changed circumstances, functional and socio-geographic, in which it operates. It must be conceded immediately that although we shall be discussing reorganization in rational efficiency terms around the notion that municipalities should be designed in relation to their functions, efficiency is not in practice the only objective in local government reorganization. In highly politicized central-local systems, party political considerations inevitably play a part simply because new boundaries always mean new majorities of voters. Because scale is related to capacity it may also mean different central-local relationships as well.

Central governments may have a stake in municipal reform in other respects as well. In the first place it may be hypothesized that all senior governments will seek, if they can, to reduce their management load by reducing the number of organizations with whom they deal. This is why central governments always prefer to deal with one pressure group which monopolizes the interest concerned rather than many. Of course, they can never achieve the same one-to-one relationship with local government by definition, but simply reducing the number must always be to some extent advantageous for the centre. Secondly, the centre may wish to see larger on average units of local government for financial reasons. One of the key characteristics of all Western local government systems is that despite the acquisition of many new, often expensive, local tasks, the local tax base has not been expanded *pari-passu*. In short, local government throughout the West has acquired a dynamic functional load but an undynamic fiscal base. This is the so-called local 'resource squeeze' problem.[1] By enlarging local government units part of this problem can be met since larger units are more likely to be financially effective despite an inelastic tax base. An enlarged system of local government may also help to simplify what may have become, because of the local resource squeeze, a highly complex system of central grants in aid.

Returning to the rational-functional cause of local government reorganization, there are usually two basic changes in the context in which local government operates that prompt enlargement. The first may be called the *socio-geographic* and it is associated with what has been called the second urban revolution that arises following the industrialization process, ie., suburbanization. The effect of suburbanization is for the urban settlements to outgrow the boundaries of the local government system. The local government structure thus becomes *underbounded* with consequent inefficiencies arising because of externalities.

The second cause of local government boundary change arises from the steady growth of local government responsibilities in, for example, the SHEW group of services (*s*ocial, *h*ealth, *e*ducation and *w*elfare). Such services often entail large items of fixed capital such as schools, residential homes, clinics and hospitals. In order to justify the cost of such capital items a large throughput of consumers is necessary. That is to say, larger base populations for local government may be required if the SHEW services are to be provided efficiently.

It will be seen immediately that both sources of change in local government boundaries point to larger jurisdictions - the socio-geographic to larger *territorial* units, the functional to larger *populations*. In short, there has been a widespread movement throughout Western states, with some exceptions, to enlarge their local government units.[2] So far we have assumed urban local government, but there has also been a perception among Western states that rural local government needed reform as well. Not because of underboundedness but, rather, because of rural depopulation, itself one of the consequences of the urban revolution mentioned earlier. Thus the rural case for local government boundary change is much more closely linked to the need to achieve economies of scale and financial effectiveness. The aim is often the egalitarian objective of seeking to ensure that rural local government can match urban government in its functional capacity so as to equalize the life chances of country and town dwellers.

Table 1: Total number of local authorities 1950 and 1992

COUNTRY	1950	1992	Evolution
Austria	3999	2301	-1698(-42%)
Belgium	2669	(1991)589	-2080(-78%)
Bulgaria	(1949)2178	(1991)255	-1932(-88%)
Czech Rep	11051	(1991)6196	-4355(-44%)
Denmark	1387	275	-1112(-80%)
Finland	547	460	-87(-16%)
France	(1945)38814	(1990)36763	-2051(-5%)
Germany[1]	24272	8077	-16195(-67%)
Greece	5959	5922	-3783(-0.6%)
Hungary	n.a.	3109	
Iceland	229	197	-32(-14%)
Italy	7781	8100	+319(+4%)
Luxembourg	127	118	-9(-7%)
Malta		67	
Netherlands	1015	647	-368(-36%)
Norway	744	439	-305(-41%)
Poland	n.a.	2459	
Portugal	303	305	+2(+0.7%)
Slovakia	n.a.	2476	
Spain	9214	8082	-1132(-12%)
Sweden	2281	286	-1995(-87%)
Switzerland	3097	3021	-76(-2.5%)
Turkey	n.a.	2378	
United Kingdom	2028	484	-1544(-76%)

[1]West Germany only, because data from 1950 for the ex-GDR are not available. The total number of municipalities including the new Länder today, however, is 16061.

Source: Size of municipalities, efficiency and citizen participation - summary report on the situation in member states prepared by the Secretariat General, Directorate of Environment and Local Authorities, Council of Europe (CDLR/Sud (94)2).

It should be emphasized that this equalization motive has not been confined to rural local government modernization, but has been one of the motivations for all local government reorganization throughout the West. Local government, because it fragments service delivery into local units with very wide variations in resources and permits the possibility of local variation, was seen by the centralizers as being intrinsically inegalitarian.[3] Local government reorganization provided a kind of compromise in this debate; by enlarging local government units the worst inequalities of resources could be ironed out thus obviating the need for centralization. In this sense reorganization became a method for 'rescuing' local government, or, rather, adapting it to the requirements of the welfare state.

So far we have discussed reorganization in fairly sweeping terms as being nearly universal, it is now time to qualify that assertion for in some states in Europe reorganization has yet to take place, as Table 1 reveals. The table provides a summary of local government reorganization in twenty four European states from the 1950s to the late 1990s and it is the most up to date available. It will be seen that in the majority of cases some reorganization has taken place but there are a few where there has been very little change and in two cases - Italy and Portugal - the number of local government units actually increased over the period.

The point to note about Table 1 is that the wide variation in the extent of reorganization (last column) seems to be linked to the type of local government system in each country. Thus for those countries which, broadly speaking, follow what may be called the *napoleonic* or *fused hierarchy* system of central-local relations (Belgium, France, Greece, Italy, Luxembourg, Portugal and Spain) each with prefects and central field services, the extent of change is relatively small, whereas in those countries which, again broadly speaking, follow a *split* hierarchy system, with no prefects, few field services, and relatively autonomous local governments, the extent of reorganization is much greater. In the napoleonic group the mean of the percentage extent of change is 5, whereas it is 41.5 for the split hierarchy group.

These results must be treated with caution because they omit a third of the countries listed in Table 1, most of which enjoy a mixed split and fused hierarchy system of central-local relations. But it is possible to claim, however, that the degree of difference between the fused and split groups which is of the order of eight does *suggest* that institutional form may be one of the factors at work. That assumption might be strengthened if it was possible to calculate a mean for the omitted group but since five of them (out of seven) do not provide a percentage change such a calculation is not possible.

The difference in the extent of reorganization revealed in Table 1 and the most obvious explanation is that the functional and geographical pressures on napoleonic systems for enlargement are simply less because economies of scale and externality problems can be accommodated since the central field services are not tied to local government boundaries. Yet the sheer extent of the disparity also suggests that there may be additional factors at work. One of these may be that precisely because municipalities in a napoleonic system are primarily representative institutions they are better able to resist reorganization, for, unlike their counterparts on non-napoleonic systems, they are not the prisoners of a system which sees local government as essentially a service provider. To be deficient in service terms for a non-napoleonic municipality is a serious weakness. For a commune in a napoleonic system, however, functional incapacity is largely irrelevant and that irrelevance presents a formidable barrier to reorganization since it destroys the central case for change. That resistance to change is likely to be re-enforced by a comparable resistance to reorganization among the central civil servants who man the field services. Making local government functionally appropriate would put their very existence into question. In those napoleonic states, like France, where local leaders are able to 'colonize' the centre via the *cumul des mandats* tradition, resistance to local government reorganization is further re-enforced since local leaders can actually control central policy at the very heart of the central decision making process and thus nip in the bud any move to put their local base at risk.

To sum up, the conclusions to be drawn from Table 1 are that in terms of local government reorganization we may perceive a North-South divide in Western Europe and it is in the North where local governments' relationship to the centre is more detached and where municipalities exercize greater formal autonomy which, paradoxically, seems to make them more susceptible to centrally imposed change.

We must now return to our main theme which is, on what basis has local government reorganization been undertaken and we begin with the *socio-geographic* factor. It will be remembered that this factor is concerned with reflecting the reality of the modern suburbanized community in the local government structure. If a city expands by suburbanization, its government, so the case for this change runs, should expand with it so as to capture the externalities so ensuring maximum efficiency in the operation of the territorially-determined services - planning, transportation, highways, and traffic management. The new suburbanized city should be matched by a unified government - 'one city, one government'.

We now come to the second major cause of enlargement, namely, the increase in the service responsibilities of local government, and in particular the acquisition by local government of the SHEW services which involve heavy investment in high cost institutions such as hospitals, schools and technical colleges. It will be remembered that the assumption in this case is that economies of scale were being lost because the existing units of local government were too small. By enlarging local government populations the 'throughput' for the said expensive institutions could be increased thus lowering average costs.

The economies of scale case also comprises two further elements, the first is that where enlargement involves consolidation (two former local governments become one) duplication is overcome. The second extra element in the economies of scale argument is that increasing scale may lower the cost of inputs because of bulk purchase discounts.

At this point in the discussion it is important to note that undoubtedly one of the influences on local government modernization in Europe has to do not with the merits of theories but with intellectual fashion. Institutional enlargement in its broadest sense has formed a kind of *zeitgeist* in the West during the latter half of the twentieth century and may be seen operating just as clearly in relation to the formation of the EC as it is in the 'merger mania' of private industry and the emergence of international conglomerates. It is also evident in general discussions about the 'shrinking' of the world in terms of communications and the expansion in international trade and discourse. Even the nation state itself has come under fire from the influence of the enlargement *zeitgeist*, it is therefore hardly surprising that it has influenced discussion of the appropriateness of local governments. Surveying the post-war period Marleen Brans has noted,

> 'The notion that large scale organization was beneficial and widespread throughout non-industrialized as well as industrialized nations, and was applied to a wide range of organizations, national and local, government and private public and semi-public enterprises. Although the alleged relationships between organizational size, structure and performance, was poorly corroborated by empirical research, it emerges in most reorganizations as a core reason for mergers and redrafting boundaries'.[4]

II. Modernization and Service Efficiency: An Assessment

(I) *Problems in Applying the Theory*: The reference in the Brans quotation above to the poor corroboration of the benefits of increased scale is an appropriate starting point for our own discussion of the application of the modernization thesis to local government reorganization and we begin with the economies of scale claim. Bran's assertion, as we shall see, is amply confirmed, but before we come to empirical verification it is necessary to emphasize, first, that structural change of the kind meant by reorganization is not the only available response to the problems of scale and externalities we have so far discussed. It is possible to achieve the necessary population and territorial scale simply by adjacent local governments cooperating in joint endeavours. Examples of such arrangements are common in many European countries.

Another response to local government modernization other than structural change is for the centre to render the smaller local units functionally more capable by making increased financial grants to them. A final functional strategy where there is a two-tier system of local government is to move functional responsibility from the smaller lower-tier units to the larger upper-tier units. In more recent times the focus of modernization has sometimes shifted to the technical and policy arena. In the search for efficiency there have been, for example, widespread experiments with new managerial forms, and most emphatic of all, outright privatisation.

To return to our discussion of the economies of scale theory, it must be emphasized at this stage that we will confine our discussion to the strict definition of the theory in terms of measurable inputs and outputs. Some who discuss economies of scale in local government often mean *financial effectiveness* which is the argument that, below a certain scale, local governments are unable to undertake certain activities because to do so would be an insupportable drain on their resources. Scale, in short, may be essential for modern local government in order to ensure that the financial base is sufficient to provide services at the range that may be essential in order to meet modern living standards. There can be no doubt that this could be a very real problem and we have already discussed the crucial importance of scale in relation to minimum population thresholds. Yet there is a problem lurking in the threshold claim, for scale can sometimes become a 'tail wagging the dog' phenomenon. This problem seems to be derived from the influence that professional specialized bureaucrats can have in specifying the various attributes that an acceptable service should have. The danger is that the ability of the local government to afford the cost of some marginal, but professionally determined, addition to a service becomes in effect the criterion for determining the scale of the whole local government.

If threshold arguments are to count some *judgement* is required which distinguishes between the core functions of a service and those marginal activities which current fashion within the profession which operates the services deems to be the minimum for an acceptable service. To put the matter bluntly, a secondary education service, for example, is primarily concerned with teachers, buildings and equipment that are directly related to the teaching of a core curriculum. There may be many essentially ancillary activities that have accrued to these core functions and all are no doubt worthy and should be taken into consideration, but they should not be the *determinants* of the appropriate scale of an education authority.

Another popular misinterpretation of the economies of scale argument in local government reform is to argue that scale is essential in order to attract, by higher salaries and more extended career structures, the best technicians and bureaucrats. A similar argument is sometimes made about elected members. Only the largest local governments, so this claim runs, will attract elected members of the necessary high calibre. The problem with accepting this claim is not that higher salaries and wider career structures do not attract better bureaucrats or that larger local governments do not attract superior politicians. No, both changes may occur with increased scale, but the problem is that larger units *require* better politicians and bureaucrats, since the management load is greater. So the question is not simply one of increasing scale in order to improve quality, but whether the quality of personnel so attracted is greater than that necessary to run the enlarged local government. If it is not, then no gain has been made by increasing scale.

It must be emphasized that the economies of scale theory is derived from the neo-classical economic theory of the firm and it is essentially a logical axiom and not an observed fact. In other words, it is a logical inevitability derived from a set of fixed assumptions. If a piece of fixed capital is used to its limit in production the average unit cost of that production will fall until the limit of the fixed capital is reached. The central assumption to note is that of manufacturing production, so absolute uniformity of output is essential to the theory.

The first difficulty in the application of the economies of scale theory to local government is centred on the character of the 'outputs' of the system. In the first place, many of these outputs may lack any uniformity.

Yet, as we have noted, if they are not uniform then the theory collapses. We repeat, absolute uniformity of output is the *sine qua non* of the theory. The second output problem concerns their measurability - we need to be able to render outputs in some form as to make comparability possible. Again for many local government outputs this is impossible without severe distortion. Finally, some local government outputs may not even be susceptible to definition, let alone comparability or measurement. What, for example, is the output of an education system?, a psychiatric counselling service?, or a domiciliary health visitor? It is true that some local government outputs are comparable to manufacturing production from which the economies of scale theory is derived. Sewage treatment, for example, or aspects of road construction. It may even be possible to arrive at *uniform measurable* and *definable* outputs for refuse collection and disposal. But the activities of normal local government that are comparable to the manufacturing process are not numerous and neither are they typical. They cannot in any sense be seen as forming the majority of most local government system's service range and that thought leads us to the next set of problems associated with the economies of scale theory and these relate to the question of service priority. It is more than likely, for example, that each major service has a different scale optimum. The scale that may be of decisive benefit for the efficient building of housing may produce *dis*economies for infant welfare clinics or street lighting. In short, the essence of local government is that it is *general* government. Local government, in other words, is always multifarious government and it follows that there cannot be one optimum scale for a given local government. If a scale optimum is to be contrived there has to be some reconciliation between the various major service scale optima. But which major functions are to be included and shall each carry equal weight? At this critical point in the discussion the economies of scale theory has little to offer.

In short, the only relevant way of applying economies of scale evidence in the determination of local government boundaries is some form of balancing or 'netting' formula and this implies some value judgement. It is irrelevant to claim that economies of scale are operating because selected services reveal it to be, unless one is also prepared to say that such services are so overwhelmingly important that their scale requirements ought to determine boundaries. Thus it is unhelpful to suggest that no more can be said on the matter because the evidence 'can point either way',[5] for different services. The crucial question has to be faced: 'which way is to count?'

This immensely difficult problem and its solution requires, as a bare minimum, that those services which are to count in the determination of boundaries are explicitly stated at the outset and that initial choice is adhered to whatever the evidence. A much more drastic resolution of the problem is to recognize the limitations of the theory when matched with the complexities of reality to tell us anything precise about scale and boundaries and therefore to ignore it. That is not to say that scale is of no consequence. Moreover in many European countries, because the largely medieval structure of local government had lingered on into the industrial era such that many very small local governments were simply incapable of performing their alloted tasks, enlargement was an urgent necessity. Precise scale disiderata was not the ruling consideration so much as creating local governments that bore some relation to the needs - both geographical and functional - of modern living. These needs clearly pointed to some sort of *minimum* scale and this brings us to the most important contribution that the economies of scale theory can make for there should obviously be some minimum scale threshold below which, for a given basket of services, no local government should fall. Helping to determine what that threshold should be seems to be the most important contribution the theory can make to local government reorganization. It could also play a part in determining the distribution of functions where there are two levels of local government - county (or equivalent) and municipality - as in Northern Europe. The key point which cannot be over-emphasized is that the irrefutability of the economies of scale theory (derived from its axiomatic status) does not in fact seem to occur empirically for a very wide range of services *where it clearly should*. In other words the failure of the theory rests not simply on a refutation of its validity of the kind we have just undertaken but is reflected in the fact that it does not seem to occur in practice. Most systematic studies of economies of scale reveal this.[6]

Whether the economies of scale theory can, or should, play a part in determining an upper threshold is not easy to determine simply because the scale of modern settlement patterns create vast urban agglomerations' in most Western states - Paris, London, Tokyo. And in such agglomerations there is usually a good case, irrespective of economies of scale, for some area-wide collective activity to cope with externalities because of the underlying socio-economic unity of such agglomerations. It is of some interest that in some large European urban agglomerations it has been found necessary to create sub-municipal councils. In this way the negative impact of scale on popular participation can be reduced.[7]

Table 2
Size of municipalities by population - current situation (1990)

Country	Total no of munici-palities	No. of municipalities by size or population (with percentage of total)									
		less than 1000		1001-5000		5001-10000		10001-100000		100000+	
Austria	2333	602	(28.5%)	1532	(55.7%)	130	(55%)	54	(2.7%)	5	(0.2%)
Belgium	589	1	(0.2%)	101	(17.1%)	171	(29%)	303	(52.3%)	3	(1.4%)
Bulgaria	255	0	(0%)	21	(8.2%)	55	(21.9%)	162	(63.6%)	16	(6.3%)
Czech Rep.	6196	4947	(79.8%)	986	(15.9%)	131	(2.1%)	130	(4.2%)	4	(1.4%)
Denmark	275	0	(0%)	19	(7%)	121	(44%)	131	(47.6%)	4	(1.4%)
Finland	455	22	(4.9%)	203	(44.6%)	120	(26.3%)	104	(22.9%)	6	(1.3%)
France	36551	28183	(77.1%)	5629	(13.1%)	898	(2.5%)	905	(2.2%)	36	(0.1%)
Germany	16061	8602	(53.6%)	4884	(30.4%)	1444	(7.1%)	1247	(8.4%)	84	(0.5%)
Greece	5922	4704	(79.4%)	1021	(17.3%)	74	(1.3%)	115	(1.9%)	8	(0.1%)
Hungary	3109	1688	(54.3%)	1152	(37.1%)	130	(4.2%)	129	(4.1%)	10	(0.3%)
Iceland	197	164	(83.3%)	26	(13.2%)	3	(1.5%)	3	(1.5%)	1	(0.5%)
Italy	8101	1942	(23.9%)	3974	(49%)	1150	(14.2%)	984	(12.2%)	51	(0.7%)
Luxembourg	118	60	(51%)	48	(41%)	5	(5%)	4	(3%)		
Malta	67	7	(11%)	30	(45%)	19	(28%)	11	(16%)		
Netherlands	647	1	(0.2%)	71	(11%)	179	(27.5%)	3378	(58.4%)	18	(2.9%)
Norway	439	17	(3.9%)	230	(52.4%)	94	(21.4%)	95	(21.5%)	3	(0.7%)
Poland	2465	0	(0%)	684	(27.7%)	4165	(47.3%)	573	(23.3%)	43	(1.7%)
Portugal	305	1	(0.3%)	25	(8.2%)	76	(25%)	180	(59%)	23	(7.5%)
Slovakia	2746	1859	(67.7%)	765	(27.9%)	60	(1.8%)	70	(2.5%)	2	(0.1%)
Spain	8086	4902	(60.6%)	2070	(25.6%)	519	(6.4%)	540	(6.7%)	55	(0.7%)
Sweden	286	0	(0%)	9	(3.1%)	55	(19.2%)	211	(73.8%)	11	(3.9%)
Switzerland	3021	1799	(59.5%)	953	(31.5%)	159	(5.3%)	105	(3.5%)	5	(0.2%)
Turkey	2378	2	(0.1%)	1886	—	—	(79.3%)	409	(17.2%)	91	(3.4%)
United Kingdom	484										

No breakdown by size of population is available, but all local authorities in the UK have more than 10,000 inhabitants

Source: The size of municipalities, efficiency, and citizen participation - summary report on the situation in member states prepared by the Secretariat General, Directorate of Environment and Local Authorities, Council of Europe (CDLR/Bud (94)2).

One final point needs to be made. When we look at the European state as it exists today we are confronted with enormous variations in the scale of their local government systems. This is clearly brought about in Table 2. Even if we bear in mind the differing historical conditions in each country and the variations in the allocation of functions between centre and locality, the economies of scale theory does not appear to have had much influence in practice on local government design. Local governments throughout much of Europe remain relatively small. In ten states listed in Table 2, for example, more than half the municipalities have less than 1.000 inhabitants and in three states the percentage under 1,000 inhabitants is in excess of 80. The considerable *variation* in the population size of municipalities throughout Europe that Table 2 reveals also reminds us that despite the importance that scale arguments may have played in the modernization process in some countries, ultimately functional capacity rarely seems to have been the primary consideration in local government design. This is obviously the case for the napoleonic group for the reasons we have discussed, but relative smallness and a wide population range are evident in the non-napoleonic countries listed in Table 2 and this suggests that unit scale, or design, is likely to have been determined not by abstract scale arguments but, rather, by the pre-existence of self-conscious sub-national geographical communities. Local government may have an undoubted functional logic in the modern state,[8] as we have seen, but it does not follow that functional considerations will always determine local government scale or design. That local government may also be a convenient service-provider is doubtless important, but it will always be to some extent a secondary consideration to the existence of definable sub-national communities. In other words, we are forced to conclude that local government is primarily a *political* entity.

There remains a further though rather less important problem with the economies of scale theory and that is, what do we mean by scale? For the sake of convenience the economies of scale theory has usually been tested in terms of population. But there are obvious deficiencies with such an assumption. Some quite small cities in population terms can be very rich in terms of financial resources. Also, the settlement pattern within a local government area must also have a bearing on outputs so that it may be misleading to assume that a densely populated city is the same in terms of its functional operation as a local government that embraces a largely rural tract simply because they both have the same total population.

Another problem that arises when the economies of scale argument for enlargement is used indiscriminately is that it tends to ignore the fact that the cost curve in the economies of scale theory is 'U' shaped. That is to say, while it is true that average costs decline as production increases on the downward slope of the curve, they just as inexorably increase during the upward slope of the curve. In other words, economies are generated at fairly precise levels of output. It follows that before the economies of scale claim can be made in relation to a given local government the shape of the curve needs to be known. What is unacceptable is the claim that economies of scale are potentially exploitable for any given local government irrespective of its size, rather as if economies of scale was a kind of medicine for a sick patient.

III. Modernization and Popular Participation: An Assessment

Following on our assessment of the impact of the economies of scale thesis in relation to local government as a service provider, we must now assess the impact of the enlargement process on local governments' role in popular participation. That is to say, if enlargement fails to make local government more efficient as a service provider does it enhance its participatory role?

(I) *Democracy as diminishing function of scale.* The first point to note is that there is an undeniable link between scale and democracy. But it is a negative one; that is to say, the smaller the size of community (ie., the smaller the number of its citizens) the greater the degree of democracy in the sense that each citizen's impact on collective decision making is greater. Very small governments, then, are the most democratic of all. The problem with this relationship is that despite its undoubted veracity the impact of the individual declines very rapidly above, say, 50 persons with increasing numbers so that the difference between the impact of the individual in a group of a 1,000 and one of 100,000 is highly marginal. Democracy is a diminishing function of scale, but in the real world where almost all units of local democracy are going to exceed 100 electors, the link hardly matters.

We move on to more relevant ground in the relationship of democraticness to small scale in regard to leaders or representatives - a crucial element in all representative systems. For we can say that the smaller the entity the more likely leaders are to be responsive to voters, since they will be, by definition, more accessible.

Thirdly, we may say that the smaller the entity the greater the likelihood of social homogeneity among the citizenry thus making possible swifter and more clear-cut majorities for collective decisions. We may reasonably conclude, then, that in terms of basic participation small scale has some undoubted advantages.

Before leaving this discussion of the relationship between small scale and democracy it is very important to discuss the contrary case, for example, by Smith who has asserted that in practice,

> 'large units of decentralized government have no worse record than small ones in some aspects of political participation'.[9]

Smith himself backs up the claim in the above quotation by arguing that larger local governments may provide *more* opportunities for participation; that smaller local authorities may tend to be oligarchic and socially conformist and possibly curmudgeonly over expenditure. A system of small local governments, Smith continues, may also perpetuate greater inequality of public resources as between different local governments. Finally, states Smith, larger local governments - because they are both financially and politically stronger - can better resist central encroachment thus preserving local democracy better.

Most of Smith's claims against the assertion that democracy is a diminishing function of scale are echoed in Newton[10] who includes one or two more participatory benefits of scale, such as better communications - radio, TV and newspapers - more highly developed party systems, and a more developed pressure group system, all of which facilitate better public participation.

These claims on behalf of scale are all persuasive and to them could be added that small communities, precisely because they may be more socially homogeneous, can be oppressive and prone to elite domination. On this score there are a number of studies that confirm the existence of such problems in small communities.[11]

However, what is being discussed in these examples of the benefits of scale are not so much scale effects but *social* effects, and we are back to the problem discussed earlier about treating population size in abstract without taking account of the settlement pattern. The reason why larger local governments have better media systems, pressure group systems and party systems is because most large local governments are cities, whereas small local governments tend to lack these aids to participation because they are rural communities. As rural communities, rather than small communities, they may also be elite-prone and hostile to dissent; even, perhaps, more curmudgeonly over public expenditure than their worldly-wise big city counterparts.

What are we to make of all this contrary evidence in relation to participation? On balance, it is fair to conclude that small scale does aid participation, but not in the clear cut manner that was suggested by the three advantages cited at the beginning of this section. Also, large scale local government may be advantageous for participation where it is combined with an urban form.

(II) *Service Accessibility*. But if we switch to another aspect of participation which is not linked to voting, lobbying and governing but to *access* to public services we may also discern a clear advantage of small scale. Small, that is, in geographic rather than population terms. This link may be stated most concisely in the form that for a given mandatory public service a local government, irrespective of its size, has to provide at least one service delivery institution (SDI) that may be required for the effective delivery of the said service, such as a school, a welfare clinic, a lending library, or a sports centre. It follows that the smaller the geographic area of each local government the greater the overall density of SDI's and hence the greater the degree of access to the citizen. This form of access, although not strictly speaking a participatory form - except where it entails access to decision makers as well as services as in deconcentrated offices of the local authority - is, nonetheless, an extremely important aspect of democratic government since if the SDI is too far away from citizens then they cannot 'consume' it and in effect the service does not exist. Accessibility of this kind may be just as crucial therefore to the efficiency of a local government as scale. If the link between small scale and service accessibility is axiomatic; the reverse statement; 'the larger the jurisdiction the lower the density of SDIs', is less clear cut. But there may be grounds for inferring that SDI density does *tend* to decline with increased jurisdictional size. This is because decision makers have a good incentive not to over-provide SDI's since they both add to service costs as well as extending the management load. Local governments will tend to provide only the number of SDI's the 'market' will bear, ie., that voters will demand. Voters may not be very demanding if low density occurs at the same time that market sector SDIs - corner groceries, petrol stations, cinemas - are also decreasing in numbers, as is often the case where car ownership is increasing. Lower SDI density in such circumstances may be seen not as a consequence of a political decision of the local government but more as a 'natural' consequence of modern society.

So far we have considered participation in relation to scale and we may summarize the conclusions that arise from it as showing that small scale local government is, on balance, preferable to large scale local government. It therefore seems unlikely that the normal enlargement arising from reorganization will tend to enhance popular participation. However, such a conclusion does not exhaust the question of the relationship that enlargement raises in relation to participation.

(III) *Enlargement and Participation.* It may be said that local government is an important element of popular participation because it both offers additional opportunities to the ordinary citizen to participate in the political process, and, secondly, it offers additional opportunities for ordinary citizens to take part in their own government. Finally, local government enhances the opportunities for popular participation by providing in institutional form a method for representing the collective interests of sub-national local communities.

How far are these key systemic functions of local government affected by unit enlargement? As to the first - voting and influencing - there would seem to be little impact from enlargement except that the seat of local government will be more distant from the average citizen in larger local governments and thus the influencing role of the citizen may be correspondingly diminished. In relation to the second, participating in government, matters are different for there is no doubt that enlargement reduces the total number of elected members in the local government system as a whole, because whole local governments are likely to disappear. For the number of councillors for a given local unit will tend to be similar irrespective of scale. Thus the council for a reformed unit will rarely have as many members as served in total on the smaller units it replaced.

We now come to the representing-the-local-community role. The essential point about this participatory function of local government is that it is based on the assumption that citizens living in close proximity to each other have objective collective interests that may need defending or promoting. But it also assumes that close proximity also engenders subjective feelings of community identity such that the said citizens feel that they have more in common with each other than they do with people living beyond their community boundaries. Such feelings are the *sine qua non* of modern democratic government for they make possible the provision of collective consumption goods for all financed by compulsory payment - taxes - irrespective of benefits received.

To what extent is this role - representing sub-national communities - of local government affected by enlargement? It will be remembered from

Section I that in addition to the economies of scale argument for enlargement there was also the socio-geographic argument which posits that local government boundaries need to be extended so as to cover the true extent of the suburbanized urban centre and its hinterland thus recapturing externalities and creating more effective planning entities. But how far does the boundary have to be extended to embrace the hinterland? Mapping the true extent of the socio-economic hinterland of an urban centre is not a simple nor an easy task. It is relatively easy to define a hinterland in the abstract, it is much more difficult to apply the theory in practice and recourse has to be made to proxy indices such as percentage of the workforce commuting to the centre, public transport networks, or labour density. At all events, if all the externalities are to be internalized the new boundary may be very far-flung indeed. But in creating such large jurisdictions the sub-national community role may be lost. In other words, a city and its hinterland may be the ideal basis for a more efficient local government in objective terms, but the new area may be far too big to engender a sense of common identity. Country dwellers and suburbanites may fall well within the compass of abstract journey-to-work maps and other socio-economic linkages, but they may not share any sense of identity with those living in the central city. Quite the contrary, they may be antagonistic to the central city and that may account for the fact of their extra-city domicile. In short, the objective criteria for delimiting modern local government units have no necessary relationship with the subjective feelings of the citizens concerned. So enlargement can mean not only a critical loss for citizens in terms of participation but unstable local government units and a loss of that vital sense of local identity which underpins the political process at the local level.

As a final comment it must be emphasized that we have only been concerned with local government modernization of a structural kind and mainly seen as enlargement. But this does not in any sense exhaust local government modernization as a general topic. In short, throughout the Western world since the 1950s there have been other changes in local government which have sought to bring local government up to date including changes in the internal management structure which sought to enhance coordination and 'management by objectives'. There have also been changes in many countries of the financial management of local authorities and experiments designed to decentralize local government administration. Most recently, and perhaps most important of all, there has been the shift to privatization which reduced local government monopoly status for some functions and introduced market-style forms of management.

NOTES

[1] L.J. Sharpe (ed), *The Local Fiscal Crisis in Western Europe: Myths and Realities*, (London: Sage, 1981).

[2] For a discussion of local government reorganization around in the world in English see:

A.F. Leemans, *Changing Patterns of Local Government*, (The Hague: IULA, 1970).

D.C. Rowatt (ed), *International Handbook on Local Government Reorganization*, (Westpoint: Greenwood Press, 1980).

A.B. Gunlicks (ed), *Local Government Reform and Reorganization: An International Perspective*, (Port Washington: Kenicat Press, 1981).

B. Dente and F. Kjellberg, *The Dynamics of Institutional Change: Local Government Reorganization in Western Democracies*, (London: Sage, 1988).

G. Marcon and I. Verebelyi (eds), *New Trends in Local Government in Western and Eastern Europe*, (Brussels: International Institute of Administrative Sciences, n.d.).

A. Norton, *International Handbook of Local and Regional Government*, (Aldershot: Edward Elgar, 1994).

[3] See, for example, G. Langrod, 'Local Government and Democracy', *Public Administration*, XXXI:2, (1953).

[4] Marleens Brans, 'Theories of Local Government Reorganization: an Empirical Evaluation', *Public Administration*, 70:3, (1992), p.449.

[5] T. Travers, G. Jones and J. Burnham, *The Impact of Population Size on Local Authority Costs and Effectiveness*, (London: Rowntree Foundation, 1993), p. 53.

[6] K. Newton, 'Community Performance in Britain', *Current Sociology*, 22, (1976).

[7] F. Kjellberg, 'A Comparative View of Municipal Decentralization: Neighbourhood Democracy in Oslo and Bologna', in L J Sharpe, (ed), *Decentralist Trends in Western Democracies*, (London: Sage, 1979).

[8] L.J. Sharpe, 'Theories and Values of Local Government', *Political Studies*, XVIII:2, (1970).

[9] Smith, *Decentralization*, p. 71.

[10] K. Newton, 'Is Small Really So Beautiful? Is Big Really So Ugly?', *Political Studies*, XXX:2, (1982).

[11] See, for example, A.J. Vidich and J. Bensman, *Small Town in Mass Society*, (Princeton: Princeton University Press, 1968).

Chapter 3

THE POLITICAL DIVISION OF LABOR

An Institutionalist Approach to Authority and Community

Henrik Paul Bang
University of Aalborg

INTRODUCTION

The study of local politics in Western Europe has mostly been conducted in the light of the notion where the state is a unified entity with specific forms of territoriality and surveillance capabilities and with effective control over the means of violence (Held, 1989). Dichotomies have been established between centralism and localism (Page, 1991) "high" politics and "low" politics (Gray, 1994), etc., presuming some sort of political division of labor between central and local politics founded on the constitutional and effective domination of the state in national affairs. However, it is highly doubtful whether the sovereignty of the state can be taken as given in the analysis of the governmental and administrative dimension of local government. Authorization and localization seem to take on entirely new political meaning when binding decisions can be made and implemented for society by public or private actors operating both "above" and "below" the state level (Crook et al., 1992, Kooiman (ed.), 1993).

If society is defined minimally as a group of actors who live together and collectively seek to cope with all the existential problems associated with their group life, politics can be defined minimally as all activities oriented towards the problem of how to allocate values authoritatively for society (Easton, 1953). It is clear that, in this minimal conception, society can be both local, national, regional, international, and global in character. Politics can be comprised of governmental as well as quasi-governmental and non-governmental activities. Yet, most political analyses are still conducted on the background of the implicit aim of redelimiting and restoring the sovereignty and supremacy of the nation-state in society.

Political scientists almost always ask, "What is the proper place of governmental authority in society" (Lane, 1993: 15), even though everybody can daily witness the displacement of this public and national authority to private, quasi-public and non-national actors outside the state. They presume that authority implies the insulation of the state from civil society as a form of legitimate 'Herrschaft' which is "exercised in a manner that is rightful, justified or acceptable" (Heywood, 1994: 78). Its operations in overnational and private institutions show that it need not reveal more than that citizens under certain circumstances are wont to accept discipline and regulation (Easton, 1958, Foucault, 1972).

The actual problem for democracy, I shall believe is the welfare state argument where democratic and social order presuppose the existence of "a central steering authority that could receive and translate into action the knowledge and the impulses from the public sphere" (Habermas, 1987:356, my italic). It is far from certain that the state is, or ought to be, the institutional medium of sustaining order and securing economic and social development in society. Furthermore, it is highly doubtful that a restoring of public authority in the state would be able to meet the growing demands for political autonomy, economic flexibility and social diversity in late-modern society (Giddens, 1991, Mouffe (ed), 1992). Not because the emerging post-industrialist information society is making political authority superfluous either as a guide to steering or as a mode of coordinating social activities. Rather because it seems to require a new, more decentralized and interactive form of political authority which does not obstruct being controlled and criticized from below in the political community.

There is a built-in tendency in the welfare state model to approach the political system 'outside in' and 'top down' as a mode of administrative domination. This results from a rationalization process 'out there' on the market place and in civil society. However, as the developments in Eastern Europe should have made evident, lack of leadership and citizenship in a political system are more critical to the existence of a political system than deficits of economic effectiveness and social consensus are. Democracy and democratization in late-modernity certainly do not require that citizens surrender their power and judgment to the state. Quite to the contrary. The dispersion of state authority seems to call for new, more direct modes of political involvement in the political community. A political system can apparently go on for a long time without receiving sufficient inputs of economic facilities and social support. But it 'stops' the very moment that it cannot get authorities to allocate values for society or get most citizens to accept and acknowledge these allocations as authoritative at least most of the time (Easton, 1965b).

To me, at least, democracy and democratization become hollow terms, if non-governmental elites, sub-elites and lay-citizens in the political community are enforced by a highly hierarchical state to exchange their right to govern themselves for material comfort and social security (Bang and Dyrberg, 1993). The exercise of appropriate leadership, as the Scandinavian welfare state model has made evident, is indeed necessary to steer between concerns for economic growth and social integration (Esping-Andersen, 1991). But such leadership becomes genuinely democratic only to the extent that it contributes to the transfer of a larger share of relevant political information and control to citizens, in particular lay-citizens in the political community (Osborne (ed.) 1991, Blackburn (ed.), 1991). The crucial political issue for democracy is not how to make the welfare state more efficient, more legitimate, more service minded or more responsive in relation to citizens. It is above all how to bring political leaders and lay-citizens back into governance in a world where more and more resources are generated into the hands of non-governmental elites and sub-elites. The future of democracy and democratization will rely on our capability to reinstitute this fundamental political division of labor in society (Easton, 1953, 1965a+b).

Rather than stating the problems for democracy in terms of individual and social rights, I shall maintain that such rights can be enjoyed only if balanced relations of autonomy and dependence exist between authorities and citizens in the political system. Rights have participation in a political community as their content (cf Bernstein, 1991). What rights are appropriate for connecting the exercise of political leadership with the building of a political community of free and equal citizens that will change over time. The political community is not synonymous with the good but with the fact of partaking in a common political division of labor with others. Thus it is about time that we begin reconsidering what self-steering and participation in a political community operating under conditions of authority is all about.

The possibility of having both authority and democracy comes out most sharply in the locality, with its institutional mixes of politics and policy and with its direct relations to the everyday life of human beings in the commune, in boards and councils, in the family, at the work place, in voluntary organizations, in social movements, etc. The locality, as Tocqueville was among the first to stress, is the place where all the modern dualities of time and space, protection and participation, authority and citizenry, politics and policy, government and administration, expert institutions and lay-practices condense. It is here we most directly can experience what it means to be, or not to be, 'politically alive' as knowing and self-governing persons sharing a common political division of labor (March and Olsen, 1989). It is also here we can most clearly see how the welfare state is becoming not only to small too handle the 'big' issues in life (human rights, planet care, economic crises, disarmament, etc) but also too big to deal with the 'small' ones (community care, local political involvement, development of labor, social integration etc., Giddens, 1991).

The Devolution of the Welfare State

Scandinavian political history in the 20th century can best be described as revolving around the social democratic welfare state project for promoting popular sovereignty and combining pressures for individual autonomy and social solidarity in a world of increasing institutional differentation (Olsen (ed.), 1991b). The social democratic welfare state ideology has continuously been questioned by liberalists and marxists, who have always been very skeptical towards the idea of using state power to meet individual and social interests. For a long time the latter clinged to the dream of getting rid of the state in order to set the associated individuals free in civil society. Normally the former have normally only tolerated its coercive power to the extent that it functions to maintain individual interests in the market place. In Scandinavia, however, both camps have had to adjust their negative view of political power to the consensus democracy brought into being by the consistent efforts of social democrats who use the state as a force of emancipation (Esping-Andersen, 1985). They have simply got to accept the social democratic view of political power as a condition of both economic effectiveness and social integration in order to stand a chance of winning in the national elections. As long as a great majority of the population perceived the social democratic welfare model as being able to serve certain goals such as security, elimination of poverty, and relative equality in income and control, liberalists and socialists could not get citizens to join them in their 'flight' from political power (cf. Rold Andersen, 1991).

From the onset the aim of the Scandinavian welfare state was from the onset the gradual realization of a consensus society, which could meet the demands of liberalism as much as those of socialism. The means of obtaining it were a special fusion of pluralism and neocorporatism, as a model of institutionalizing the social struggle between work and capital through overlapping group membership combined with extensive collaboration between organizations, government and bureaus (Pedersen, et al, 1992). Centralizing the organization of labor and capital and incorporating them into public interest mediation and policy implementation, the state succeeded in making them function as the prolongation of its executive arm. A state-sponsored and state-controlled 'class-compromise' was forged, strengthening elite consensus and elite unity in policy-making (cf. Damgaard (ed.) 1986). This was believed to be able to lay down a consistent line of intervention to deal with the revolutionary claim of continuing social reorganizing without ending up negating civil liberties (cf. Donzelot, 1991, Crook et al, 1992).

The corporate 'power politics' of the social democratic welfare state combined interest group politics, elite politics, and consensusal politics to guarantee sufficient 'inputs' of demands and support and to pave the way for the provision of welfare 'outputs' by an increasingly disethicised, pragmatic and status-oriented bureaucracy. For many years this institutionalization of the distinction between individual autonomy and social solidarity as a pluralist-corporatist system created a political situation in Scandinavia where the state was no longer at stake in social relations, but stood outside them as a guarantor of their progress (Selle, 1990). Presenting itself as a force of emancipation in civil society, tied to the expansion of both economic effectiveness and social integration, the Scandinavian welfare state could momentarily passify both its liberalist and socialist opponents. On the one hand, it could act an instrument of freedom, oriented towards releasing individuals from the rigidities of tradition, safeguarding human rights, and enlarging economic opportunities. On the other hand, it could serve as a medium of social order, bringing about a peaceful socialization of the capitalist economy, reducing anarchic competition and economic waste, and protecting 'the weak' against the 'excesses' of industrialist capitalism (Esping-Andersen, 1991).

However, the more 'state-like' interest groups and parties became in their mode of functioning the more difficulties they experienced with sustaining their image as independent intermediaries between the state and the individual. Pluralism came in deeper and deeper conflict with neocorporatism. Representative democracy began to loose its identity as a guide to public decision-making, because more and more interest politics was transformed into technical policy problems to be dealt with by experts from the state, capital and labor. As a consequence the welfare state began to function more as an all-pervasive disciplinary power for the technical elimination of dysfunctions than as a pluralist model of democracy (Squires, 1990). Through the development of administrative domination and corporatist modes of surveillance and regulation, the state could penetrate deeper and deeper into every pore of society, constantly keeping in touch with the condition of market health, the commune, the family, and other socializing agents (Foucault, 1980, 1990). Hence, state power in its exercise became a much more local, comprehensive, ambiguous, differentiated, social and mundane phenomenon than was intended by the system of rights in terms of which the sovereign state was originally separated from civil society.

How could liberals and socialists tolerate the dramatic increase in state interventions in Scandinavia in the 1950's and 1960's? Liberalists could accept state growth, and even happily contribute to it, because it was addressed towards economic development, showed reverence to citizenship rights and enjoyed the support of most of the citizenry. Inversely, socialists could live with the expansion of the state's regulatory and disciplinary capabilities, because the welfare benefits and social rights implemented by its administrative and corporate institutions curbed their hopes for bringing an end to economic exploitation. However, the period of relatively unproblematic state growth came abruptly to an end with the 'youth revolution' in the late sixties and early seventies. At this date individuals and groups in the locality finally began to question the validity and democratic nature of the strong corporatist and etatist institutions produced by the 'class compromising' Scandinavian model (Olsen, 1991b). The social democratic welfare state project slowly lost credibility and had to hand over the social and political initiative to neo-conservatives and neo-liberalists whose slogans about governability and market democracy finally conquered the agenda of discussion in the 1980's (Fonsmark, 1991). Yet, in a way the nature of their debate seemed to crown the social democratic triumph, precisely because it was orientated towards problems of the state's 'ungovernability' and 'delegitimation' rather than towards a genuinely political and democratic restructuring of the social system.(Thomsen, 1992).

Both liberal and socialist critiques of the welfare state adopt the welfare state approach to politics and policy as concerned with mediating demands for economic effectiveness with needs for social consensus. They only offer little advise with respect to the personal and communitarian problem of self-government and popular sovereignty brought about by the growth in the state's disciplinary and regulatory powers. They overlook that for a long time when the welfare state was able to generate legitimacy for the democratic regime and to steer appropriately between the market and the civic culture. It was not primarily due to its 'inputs' of economic facilities and social support. It was above all the accomplished outcome of its internal functioning as an etatist and corporatist armistice achieved under the aegis of a new leadership dramatically extending the intervening powers and regulatory functions of political authority in society. This new worldly authority seemed to offer society exactly what both liberalism and socialism excluded - a positive view of political power, of political institution-building, and of political leadership enabling the emergence of a new political economy possessed of highly specific procedural techniques, entirely new instruments, and a range of novel apparatuses (cf. Foucault, 1980:104).

For a long time the temporalized and highly differentiated power of modern authority made it possible for the welfare state to function in society as if it were a universally rightful sovereign authority backed by universal rules, general procedures, comprehensive political representation and separation of power. In the end, however, it was exactly its worldliness as a corporatist-pluralist system of disciplinary subjection which made citizens aware of its limits as a means for emancipating them from the fixities of tradition and from conditions of hierarchical domination. Today, the state has once again become an issue in social relations. It can no longer hide the problem of illegitimate domination derived from the continuing corporatization and etatization of political power behind the 'neutral' language of statistics by which potent social conflicts are sought and kept in order. More and more citizens in the local political community are becoming aware that the state's constitution of public right articulated upon collective sovereignty is fundamentally conditioned by and grounded in the democratically unfounded disciplinary power of various professional and administrative agencies and policy institutions (cf. Foucault, 1980: 105).

The increased politicization of the locality indicates that the welfare state's increased 'ungovernability' and 'delegitimation' need not be a sign of crises tendencies but of a 'second wave' of steering and participation from below in the political system (Steenbergen (ed.), 1994). There are some indications in the locality that orientations are changing, in particular among young citizens, towards forms of more personal and more multilayered democratic participation for coping with those moral and existential questions of survival and being, transcendence, cooperation and personhood which have been repressed by the interest politics and emancipatory policies of the welfare state (Andersen et al, 1993).

The 'post-materialist' generations (Dalton and Kuechler, 1990) do not appear to mistake political autonomy for economic effectiveness or democratic identity for social solidarity. As a young Dane, having been an activist for almost 13 years, said in an interview about the ongoing, often violent, struggles between the police and the 'autonomous' in the inner city of Copenhagen: politics "is a life and an everyday activity. This is the difference between traditional politics and the autonomous movement. I do not go out to be political and then go home to be something else" (Politiken, 29.04.1994). As a local citizen, the activist cannot make sense of the welfare state argument that "[t]he stronger, the more active the State becomes, the more the individual increases his liberty. It is the State that sets him free" (Durkheim in Giddens (ed.) 1986: 50). He knows that being free means being able to make a political difference, which is why to him liberty as the product of state domination is simply a contradiction in terms (cf. Bang and Jacobsen, 1994).

In the local political community, there is a growing awareness of the fact that it is participation, rather than consent, which legitimates government - in particular among those individuals and groups who are not represented, and have no democratic stake, in the games that organized interests play in the negotiated economy of the Scandinavian welfare state (Andersen, et al, 1993). The abstract idea of the state as linked to the common interest has been paralyzed by the concrete operations of the corporatist-pluralist system as a set of relatively autonomous political institutions each of which is pursuing a set of very specific and extremely precise interests related to demand aggregation, issue articulation, policy implementation, outcome justification, etc. However, even this system can no longer resist the pressures from the many new global, international, regional and local institutions and agencies involved in national decision-making and implementation (Goetz and Clarke (eds.), 1993, Waever et al, 1993).

More than anything else the problem for the Scandinavian welfare state is that it has created a situation for itself where it has become almost impossible to distinguish the role of authorities in the state from the role of elites, sub-elites and lay-actors in the political community. Leaders in interest organizations, for example, function as much as state authorities chronically involved in the disciplining and regulating of citizens on the output-side than as gatekeepers, aggregating and articulating interests on the input-side. At the same time they are also engaged in reforming and reorganizing the state as participants in a complex network of privately and transnationally operating elites and sub-elites (Pedersen et al, 1992). In any case, parliament and lay-citizens seem to be those who suffer the consequences. They not only have to deal with the fact that national public policy is regularly produced by non-national public administrations. They also have to accept that public politics are often articulated by elite networks not directly related to the exercise of political leadership and lay-citizenship, at whatever level. In addition, they must be prepared to face the whole myriad of new practices or institutions that result from the mixing of authorities, elites and sub-elites within or without the state. Indeed, these clashes of governmental authorities and non-governmental elites and sub-elites "may have a liberating and equalizing potential, which we could disregard only at our peril" (Etzioni-Halevi, 1993:7). But however high this potential may be, it is no substitute for the exercise of appropriate leadership and sound lay-citizenship.

Actually, we seem to have reached the point where decisions which are made authoritatively fuse with choices which are made individually as a huge network of power and knowledge circulating freely between global, international, regional, national and local expert systems (Pedersen et al, 1992). No wonder, therefore, that the Scandinavian welfare state is experiencing severe troubles sustaining itself as a sovereign entity possessing the ability to command and administer its own territory. Scandinavian politics and policy are no longer taking place in Scandinavia only, and both government and administration are 'colonized' by a host of extraparliamentary and privately operating agencies and institutions all of which claim their share of political autonomy and identity. Political responsibilities are continuously negotiated by elites rather than calculated by public authorities; demands are voiced by non-national lobbyists operating outside the domains of national interests organizations and parties; issues are formulated by private organizations and voluntary associations outside the public parlors of parliament; decisions of national consequence are made by international corporations beyond the reach of government; and policies are implemented by transnationally operating networks working together by local authorities.

From State Politics to Life Politics

How do lay-citizens respond to this new democratic situation? In many different ways, of course, but there seems to be a marked decline of interest in representative politics in the state. However, the withdrawal from traditional interest politics may range from total political abstinence to complete immersion in a new politics of choice in the locality, comprising a variety of single-issue movements and groups together with a host of new life-style policies (Gundelach and Siune, 1992). The shift seems to be away from the state and towards 'the local in the global'. There is a growing concern with those life-political problems that the state has systematically neglected and a new awareness that their solutions will require the transfer of more power-knowledge to local and global forces and institutions.

I shall give some examples by drawing on a comprehensive survey analysis from Aalborg University concerning the attitudes of the members of Denmark's greatest union TUC (1.327.506 members) towards their work-place, union, family and state. Although the project was not designed to cope with life-political issues of autonomy and identity, it nevertheless indicates that citizens are becoming increasingly aware of politics and policy as an endogenous aspect of their everyday life. Questions of citizenship are extended far beyond the state and local government to comprise various new modes of local governance crossing established boundaries between private and public and state and civil society.

The members of TUC, we shall see, apparently want to use the workplace as a site for developing their autonomy and identity as citizens. They are aware that this involves appropriation of resources as well as molding rules and identities and of politically socializing individuals into them (cf. March and Olsen, 1994:13). However, they do not neglect the division of labor between their leaders and themselves in welfare politics and policy. 87% of the members agree, or partly agree, that unions are necessary to meet the interests of employees (Joergensen et al, 1992: 240). Furthermore, the members show great reverence for welfare benefits as such. 90% agree, or partly agree, that the health care system should be extended so that waiting lists can be avoided (Joergensen et al, 1992: 321). 91% agree, or partly agree, that old-age pension should be improved. 39% accede, or partly accede, that taxes are too high and that the public sector should be cut down (an astonishing low figure, all things considered). Only 14% think, or partly think, that child care should be limited (ibid). 85% consider it important to fight unemployment (ibid:314). However, the point is that apparently the members strongly mistrust the ability of the welfare state and its corporate-pluralist system to cope with the current problems of welfare and democracy in an appropriate manner.

Only 15% of the members agree, or partly agree, that they can get their way in relation to their union (ibid:252). The union leadership is perceived as a barrier to their own self-development and as outside the reach of substantial control 'from below'. Merely 32% are convinced, or are partly convinced, that the union leadership is acting in accordance with their interests (ibid:255). 61% think, or partly think, that the distance between their work place and their interest organization is too long (ibid). Finally, 58% want, or partly want, their local branches to decide more in relation to the central organization. This shift from organization to movement and from central to local involvement seems to come out in the voting patterns of the members:

Table 1. Regular voting grouped according to age (%)

	How do you Vote?				
Age	To the left	To the right	In the centre	I am changing between 1, 2 and 3.	I usually do not vote
20 years or less	36	10	(6)	11	38
21 - 30 years	45	26	5	10	14
31 - 40 years	63	17	6	9	5
41 - 50 years	68	15	6	8	3
51 - 60 years	69	15	7	9	1
61 years or more	61	17	(2)	20	(1)

Whereas 69% of those between 51 and 60 years vote for the social democrats and the socialist people party, only 45% between 21 and 30 years do so. Whereas 14% between 21 and 30 years do not normally vote, this is only the case with 1% of those between 51 and 60 years (Joergensen et al, 1992: 313)!! What we have here is perhaps a generational gap (rather than a cyclical phenomenon), maybe even a ticking bomb under representative democracy in Denmark.

Furthermore, even though the authors have organized the replies on the left-right axis, in actual practice it seems as if this axis is beginning to lose its importance as a criterion for political identity (cf. Andersen et al, 1993). First of all, there are many young members who do not vote at all, and those who do vote to the right do not behave as expected with respect to key-issues of privatization, free riding, the state as a supermarket, cut-downs in welfare benefits, order, the flag etc. Rather the young generations seem to shy away from all the dualisms associated with the opposed steering principles of hierarchcy, anarchy and solidarity characteristic of the battle between liberalism and socialism in their common rejection of conservative tradition and custom. For example, comparing those between 51 and 60 years with those between 21 and 30 years (that is, those who grew up with the welfare state and those who simply take welfare for granted), one finds that the 'young' rate solidarity much lower than the 'old', (in weighted figures 3,0 versus 1,5 with the average being 2,4 [the lower/the higher], Joergensen et al, 1992: 242). On the other hand, the 'young' are considerably less interested than the 'old' in security and order (2,1 versus 1,6), the spread of market mechanisms (2,6 versus 2,1), tradition and morality (2,7 versus 2,1), conscientiousness (2,3 versus 1,8), and more prosperity (2,3 versus 1,8, ibid: 338). Telling figures, which make one wonder why so many see the flight from the 'old' left as the victory of the 'new' right, when in practice lay-citizens seem to be in the process of putting both of these narratives behind them.

Inglehart's "postmaterial values" appear to be of some relevance (1990) for understanding this beginning shift of interest from state concerns with economic effectiveness and social solidarity to local anxieties with global and personal affairs. The members of TUC rate protection of the environment and improved life quality even higher than economic growth and improvement of real wages (95% and 85% versus 87% and 67%, Joergensen et al, 1993: 49). When asked what would mean the most to them, if they could choose a job freely only 33% refer to high wages as the most important incentive, whereas 77% point to a good working environment and 79% to a management showing tact and respect of their employees (ibid, 1993b: 28). But more consequential to the analysis of local governance in a democratic perspective is the fact that the growth in immaterial values indicates a growing concern with power and knowledge in the locality as intrinsic to the shaping of personal and collective autonomy and identity.

This is perhaps why the members of TUC put more confidence in management at their work place than in their own representatives. Whereas 86% say that they share, or partly share, interests in common with management at their work place (Joergensen et al, 1992:197) only 58% say that they have become, or partly have become, members of the union to be solidaric with their fellow workers (ibid:241). Obviously, the class struggle between capital and work upon which the labor union was founded has more or less lost its significance for (especially the 'young') members. Rather the members seem to perceive management as a potential ally in their struggles against the distant authorities in interest groups, parties and government:

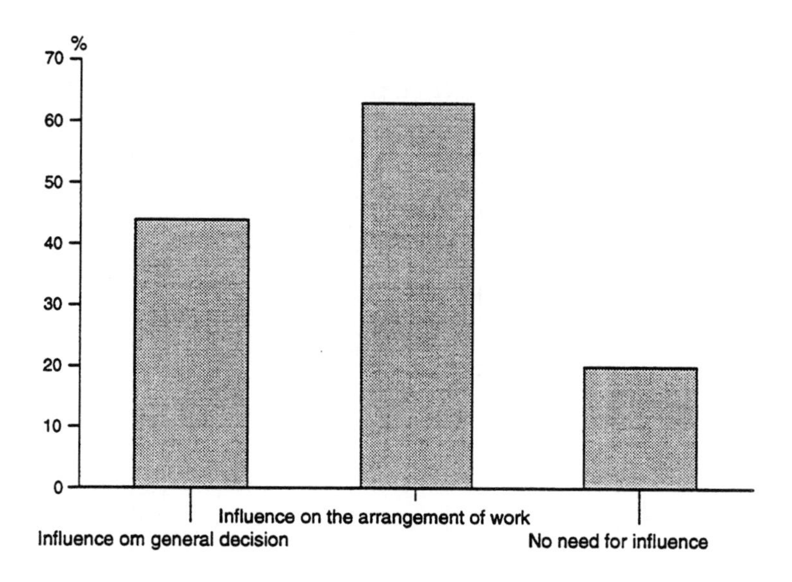

Figure 1. The members' endorsement of self-governance at the work place. (%)

The interesting thing here (Jørgensen et al, 1993b:31) is not merely the high percentage in favor of more influence but, in particular, that the members rate influence in the arrangement of work higher than influence in general decision-making. This shows today how political issues of power-knowledge are invading the work place, supplementing the capital-labor relation with one of leadership and citizenship, bringing forward a new politics of choice and self-governance 'from above' as well as 'from below'.

Indirectly, the members of TUC seem aware that their own autonomy and identity as citizens depend on "the willingness, readiness, competence, resources and capabilities of the authorities for acting upon returning information and taking follow-up measures to attain their aims" (Easton, 1965b:433). Their attitudes towards general and specific control at the work place demonstrate that life political issues, such as those relating to the structuring of their work, cannot be debated outside the scope of the strategic decisions and actions of management as an abstract system. At the same time they display a renewed sensitivity to all those questions which the institutions of modernity systematically tend to dissolve - questions of Being, new modes of reflexive cooperation, mediation of self and body, etc. (Bourdieu, 1990).

Table 2. Perceived benefits from courses/education in hours of work distributed according to sex and age (%).

	What did you get out of participating in courses/education?						
	benefit						
	in relation to work	in relation to job change	in relation to labor market	in relation to experience of community	in relation to continued learning	in relation to push in personal development	expectations not fulfilled
Sex							
women	74	9	12	46	63	47	7
men	75	16	18	33	59	33	8
Age							
20 years or less	74	(21)	29	65	66	41	(9)
21-30 years	78	17	15	30	50	40	8
31-40 years	74	13	16	39	59	43	6
41-50 years	75	9	14	45	68	39	8
51-60 years	67	7	13	46	69	38	8
61 years or more	71	(12)	(13)	50	62	30	(18)
I alt	74	12	15	40	61	40	8

What intrigues me is that the members seem aware that local self-identity is a reflexive achievement which is formed, altered, and maintained in relation to rapidly changing circumstances of decision and action. In their practice they insist on treating power-knowledge as inseparable in political action. This comes out clearly in their approval of further training during work hours (Joergensen et al, 1992: 185).

Surely, these figures communicate that the uncertainties that the members face in their daily life are less uncertainties about outcomes and preferences than they are uncertainties about the demands of identity (March and Olsen, 1994:8). They show that learning is for life in the local community of late modernity and that citizens appreciate the importance of being able to "integrate information derived from a diversity of mediated experiences with local involvements in such a way as to connect future projects with past experiences in a reasonably coherent fashion (Giddens, 1991:215). The members' interest in knowledge seems to be guided by their mutual understanding of the often confusing, uncertain, ambiguous and risky world they live in as well as by their perception of the possibility of using various life-stile patterns as a means of combating the experts systems which continuously threaten to undermine their self-identity and self-determination.

That we are dealing with a beginning change from an economy of choice to a polity of choice becomes evident from the members' description of the inverse relation between conditions for improving self-governance and the presence of a highly instrumental leadership predominantly oriented towards economic effectiveness (Joergensen et al ibid):

Table 3. Attitudes towards self-government at the work place collated with agreement as to whether "the leadership places effectiveness before all other goals"

	Should there be more self-governance at the work place?		
"Effectiveness before"	general decisions	organizing of work	no demand for increased self-governance
to a very high degree	54	72	12
to a high degree	45	66	17
partly	44	57	23
to a small degree	38	59	29
to a very small degree	44	48	31

The more the leadership commits itself to principles of rational choice the more its employees apparently commit themselves to principles of self-governance. Table 5 communicates an interesting discovery about the relation between democracy and market. It seems to manifest an implicit agreement to act appropriately in return for being treated appropriately in the authority relationship. It illuminates that lay-actors normally act in their practical consciousness to being parties to a cognitive process of interpretation about what is to constitute exemplary, 'natural' or acceptable behavior. At least, they appear to act upon identity-driven images of appropriateness more than upon economic calculations of utilities or upon the "unifying, consensus-building force of a discourse in which the participants overcome their at first subjectively biased views in favor of a rationally motivated agreement" (Habermas, 1987:314-315). They seem to recognize that political identities and rules assure neither consistency nor simplicity but are open to continuous changes 'in both directions'.

Hence, the survey study from Aalborg illuminates that lay-citizens are able to assess what rules and identities exist in institutions, which ones are dominant and what the tension between effectiveness and self-governance implies. They express an awareness of the limits of instrumental and communicative rationality and of their embedding in an institutional context with specific constellations of political rules and resources. As such they seem to confirm that generally citizens are conscious that they "are limited by the complexities of demands upon them and by the distribution and regulation of resources, competencies and organizing capacities, that is by the capability for acting appropriately" (March and Olsen, 1994: 10).

The orientations of the members, in particular of its young members, may reflect a silent rebellion against the tendencies of state authorities and other leaders to turn them into mere clients or ciphers in an overly disciplinary and self-regulating system. They indicate that they are willing and ready to fight political domination and to take active responsibility for their own societalization as knowledgeable and capable human beings reclaiming their political identity and sovereignty. They seem to turn against their organizational leadership, because they perceive it as a part of the state hierarchy. Instead they conceive their management at the work place as a potential ally for developing alternative modes of governance. This could obviously be a dangerous course for democracy, closing the eyes of citizens to both the need for authorities, at whatever level, and to the fact that management has been provided the right by the state to discipline and govern them without any legitimate props to their obeissance. Yet, it may also be regarded as a new possibility for expanding citizenship to comprise the work-place as open to constructive interpretation, criticism and justification of political rules and identities.

The Aalborg study thus communicates that the locality, in its widest sense, is well suited for activating citizens, organizing their practical knowledge of what has to be done, reducing their dependence on services, and providing them with a strong autonomous element through control of resources (Chanan, 1992). It confirms that citizens in the locality have fully absorbed the new politics of choice in 'postmodern' society, with its slogans of 'universal human rights', 'global planet care', 'international disarmament', 'personal is political', 'autonomy before rights' or 'better self-responsible than subordinate' (Giddens, 1991). It illuminates that organized opinion cannot replace the genuinely popular one and that there is a profound need in the locality for enhancing lay-citizenship control of rules and resources in the political community.

The Return of the Political Community

Making politics and policy a part of everyday life may be the first step in establishing new and more reciprocal relations of governance between experts and lay-actors (McLennan and Sayers (eds.) 1991). The 'old' steering models of state ('hierarchy'), market ('anarchy') and civil society ('solidarity') seem peculiarly outdated for studying the new mixes of leadership and citizenship in the locality. They should "be supplemented with (or replaced by) a perspective that sees the polity as a community of rules, norms and institutions" (March and Olsen, 1989: 171). On the one hand, we should try to tie the new issue politics and life-policies of citizens to the development of both the practice of self-governance and a sense of mutual identification of one's common capacities for decision and action. On the other hand, we should attempt to restore the interest politics and emancipatory policies of the welfare state within a non-hierarchical approach to "authorities in political systems [as] differentiated by the special capabilities that they possess to mobilize the resources and energies of the members of the system" (Easton, 1965a:54, cf Burchell et all, 1991). Effective services require democratic cooperation, and life-political issues regarding our political responsibilities towards nature, the unborn, our body, and each other as citizens cannot be debated outside the scope of all those abstract expert systems which confront us in our everyday life. Information drawn from various kinds of expertise in the environment is central to the forging of one's autonomy and identity as a citizen in the local political community (Giddens, 1990, 1991).

Then how is local political analysis then to cope with the problem of combining control 'from below' with steering 'from above'? First of all by acknowledging that politics and policy manifest a power and knowledge of their own, not derivable from either the logic of consequentiality on the market place or the power of persuasion in the normative culture (Easton, 1993, March and Olsen, 1994). Political institutions seem more directly connected with the development of political autonomy and democratic identity than with aggregating individual preferences and integrating social norms. They should therefore be explicitly designed to meet concerns for appropriate steering and popular control, since the exercise of competent leadership and citizenship are the basis for establishing a politically balanced mediation of economic effectiveness and social solidarity in society. Policy and politics in the welfare state have been much too concerned with what politicians, bureaucrats and other experts say and do. As a result it has been neglected that citizens are necessarily involved in, and responsible for, the structure of the political system as participants in a political division of labor held together by authority (Schwartz, 1988).

Normally, the opposition between libertarians and communitarians in our discipline is described as one between the private individual and her rights and the social community and its good (Mulhall and Swift, 1992). The political community, however, does not have its basis in either an atomistic, abstract, and egotistic individual as the subject of rights or a collectivistic and equally abstract solidaric humanity as the object of good. Rather it springs from the recognition and acceptance of political authority as a condition of transforming a state dominated community into a genuinely democratic one of free and equal citizens. This is also why individual and social rights cannot occur outside the political system. "[T]he polity embodies a political community and the identities and capabilities of individuals cannot be seen as established apart from, or prior to, their membership and position in the community" (March and Olsen, 1989:161).

Citizenship does not, and cannot, mean freedom from politics and policy, as liberalism will have it. It springs from the transformative capacity of citizens to participate in a common political structure and the set of decision-making processes in a knowledgeable fashion, however tight or loose their internal political relations may be (Bang, 1987a+b). If citizens abandon their right to govern themselves in the political community for material comfort and social security they are simply no longer citizens (Bang and Dyrberg, 1993). Every citizen bears some responsibility for the way collective decisions are made and implemented for society. Not the same degree of responsibility, of course, since most citizens are not directly involved in the production of authoritative 'outputs' or in elite negotiations. But some responsibility even the most oppressed and marginalized citizen does have otherwise s/he simply would not be a citizen in the first place (Clark (ed.), 1994).

The political community is a phenomenon which is neglected in the modern traditions. However, it has always been 'there' inside the political system as evidence of a classical mode of being which resides in nothing outside the political division of labor between citizens and authorities in the political system. As Perikles was the first to emphasize, the political community does not exist as given, whether by God(s), a certain class or race or sex. It comes into existence, because "[w]e give our obedience to those whom we put in positions of authority, and we obey the laws themselves, especially those which are for the protection of the oppressed, and those unwritten laws which it is an acknowledged shame to break" (in Held, 1987: 16). This ancient idea of Polis as the basis of both leadership and citizenship seems particularly relevant to revive today. It provides us with the possibility of helping to establish new central-local relations founded on the mutual acknowledgement and acceptance of political authority as a power-knowledge which, as Rousseau put it, is "one and simple, and cannot be divided without being destroyed" (1762, (1973): 2-63). It offers a way of replacing hierarchy by the recognition and acceptance of the fact that we cannot all be authorities who are directly responsible for the day-to-day articulation and implementation of binding decisions and actions for society.

The roles of authorities and non-authorities should be kept distinct, at least if we wish to contribute to the formation of a genuinely interactive democracy in which decentralization, modernization, deregulation, and privatization do not just manifest a transfer of effective power from state elites to non-state elites. It is far from obvious that the new strategies of user influence and spontaneous social solidarities will always contribute to enlarging the capacities of lay-citizens to 'make a difference'. In actual practice, their intended or unintended outcomes may just as well be to expand the control of authorities in local affairs or to increase the relative autonomy of non-governmental elites and subelites as against both authorities and lay-citizens.

Where neo-liberalists and neo-socialists tie the lack of active citizenship in the welfare state to the suppression of individual preferences and the reification of social norms, I would rather relate it to attempts at colonizing the way authority functions as a communicated message in the practical consciousness of human beings. This problem cannot be appropriately dealt with in terms of either neo-liberalism with its rational individual actor or neo-socialism with its harmonious social structure. The political dilemma for lay-citizens is not that the welfare state undermines private incentives and no longer facilitates the translations of individual preferences into observed political outcomes as an instrument of aggregation, as those from the New Right will have it (Buchanan, 1987:339). Nor is it that this state stands opposed to social integration as a systematically integrated sphere of action distorting the communicative attempts of associated individuals to reach a normative agreement in dialogue, as those from the New Left claim (Habermas, 1982:265). More than anything else the dilemma concerns the monopolizing of allocative and authoritative resources by authorities, elites and sub-elites, preventing the lay-citizen from regularly participating and chronically in the control of politics and policy for the sake of developing her personal autonomy and identity in cooperation with fellow-citizens in the political community (cf. Mclellan and Sayers, 1991).

Consequently we should operate with three basic constituents of the political system:

Political authorities are "occupants of authority roles as elders, paramount chiefs, executives, legislators, judges, administrators, councilors, monarchs, and the like. At times, they may be occupants of highly differentiated roles as in modern political systems; but in many less specialized societies particularly, the occupants may not perform in a specifically political role" (Easton, 1965B: 212). In any case, however, they "may be said to conform to the following criteria. They must engage in the daily affairs of a political system; they must be recognized by most members as having the responsibility for these matters; and their actions must be accepted as binding most of the time by most of the members as long as they act within the limits of their role" (ibid).

The political regime "refers to the general matrix of regularized expectations within the limits of which political actions are usually considered authoritative, regardless of how or where these expectations may be expressed" (ibid:193-194). It consists of "values (goals and principles), norms, and structure authority. The values serve as broad limits with regard to what can be taken for granted in the guidance of day-to-day policy without violating deep feelings or important segments of the community. The norms specify the kinds of procedures that are expected and acceptable in the processing and implementation of demands. The structures of authority designate the formal and informal patterns in which power is distributed and organized with regard to the authoritative making and implementing of decisions" (ibid:193).

The political community concerns "that aspect of a political system that consists of its members seen as a group of persons....who are drawn together by the fact that they participate in a common structure and set of processes, however tight or loose the ties may be" (ibid:177). That is to say, "it does not matter whether the members form a community in the sociological sense of a group of members who have a sense of community or a set common traditions" (ibid). In fact, "they may well have different cultures and traditions or they may be entirely separate nationalities" (ibid). "But regardless of the degree of cohesion among the members of the system, as long as they are part of the same political system, they cannot escape sharing in or being linked by a common division of political labor. This forms the structural connection among the members of the system that gives minimal linkage to political activities that might otherwise be isolated or independent" (ibid:178).

In local political analysis, the notion of local community action has recently gained new prominence as a catch-all category for discussing the new notions of issue networks, policy communities, the negotiated economy, user-influence, direct democracy, etc. (Chanan, 1992, Kooiman (ed.), 1993). As defined here, however, the local political community is not an emerging 'third sector' or 'quasi-public sphere' which has some degree of visible public presence as distinct from, and taking place outside of, the private life of individuals and the public life of the state. It is simply an aspect of the political system penetrating into even the remotest corner of society. The democratic community is not the place for private individuals who "acquisce in the coercion of the state, of politics, only if the ultimate constitutional 'exchange' furthers their interest", as the New Right proposes (Buchanan, 1987: 338). Nor is it simply "a non-state sphere comprising a plurality of public spheres" (Keane, 1988: 14) as those from the New Left put it. The political community is a necessarily disturbing influence in the present, as manifestations of a power-knowledge which comes from everywhere and in principle belongs to everybody.

The political community shows that "[a] specific regime at any moment in time will be the product of the accommodation among the pressures for new goals, rules, or structures stimulated by social change and the limitations imposed by existing conventions and practices" (ibid:194). Political authorities are necessarily tied together with the political community by the fact that they could not function as direct producers of political 'outputs' unless citizens obey and consider themselves bound by them. Inversely, the day-to-day recognition and acceptance of political authority in the political community manifest that political authorities will always have some autonomy in the political system. This is because citizens, however powerful some or all of them may be, cannot themselves make authoritative decisions and actions.

Rather than studying local governance solely on the background of the juridical edifice of sovereignty, i.e. the political authorities, the regime, and the formal apparatuses and steering ideologies that accompany them, we should consequently begin assessing the limited field of juridical sovereignty and state institutions in the light of the wider field of the diversified tactics and strategies of power that underlie the life-politics and different policy-styles of actors or collectivities in the political community (Foucault, 1979, 1980). This would reveal that the roots of modern democracy are not merely 'right', that is, regime principles and rules and the complex of apparatuses, institutions and regulations of authority responsible for their day-to-day application. More fundamental than right is "sharing political processes, participating in interdependent political roles, and partaking in the same communication network" (Easton, 1965b: 327). The development of a sense of community (political solidarity) cannot replace the development of such participation. Only by enhancing communal steering and participation can we become able to cope with the "difficult problem of balancing the undoubted advantages, even necessity, of institutional autonomy with the risks that such autonomy will make popular control difficult or impossible" (March and Olsen, 1989: 166).

The notion of political community comprises (1) the distinctive class of actions in which human beings are engaged as a group of persons with interlocking roles and practices; (2) the network of concepts which inform these persons and endow them with meaning and a mutual identity; (3) the projects and purposes which are generated and developed by them in and through their relationship with the political authorities; (4) the resources (whether material or immaterial) which are utilized by them to gain autonomy in their performance of issue politics and life-style policies ; and (5) the political and environmental conditions and constraints under which their communal identities, the resources they require, and the power relations they engender are shaped and directed in numerous, sometimes internally inconsistent, ways (cf. Reed, 1985: 120).

The return of the political community will probably force us to swallow a lot of 'old camels' in local political research. We can no longer take it for granted that citizens are either a permanent thorn in the side of political authorities or innocent victims to their arbitrary 'will to power' (Held, 1987, Keane, 1988). Every citizen, however 'marginalized' s/he might appear to be, must be thought of as bearing some influence upon why things are as they are, or happen as they do, in the political system. Nor can we presume any longer that "[c]itizens who live together under democratic conditions must be obliged to submit themselves to a highly centralized authority, without the rational domination of which they would fall into confusion and disorder (Keane: 1988: 237). If citizens fall into confusion and disorder, the entire system will necessarily do the same.

Even the choice to live privately in splendid isolation from politics and policy will reflect a life-decision made by a citizen within the political community. Citizens cannot run away from their political responsibility. Furthermore, without relatively autonomous citizens there simply would be no reasons for having authorities. Citizens are those who pose the demands and issues that authorities are called upon to convert into binding decisions and actions and failure to cope with them may arouse citizens to active resistance and action. However, not even the strongest class or elite citizens are authorities but respond to authorities, as receivers of their communicated messages. This is important to stress, since it illuminates that it is the awareness of authority which enables citizens to make demands, issues and evaluations about the way it should be exercised. How they respond then measures the extent to which the outputs of authorities have proved effective or not. The authorities may, through incompetence or lack of resources, fail to decide and act upon their responses. But without 'feedback' of some information to the authorities about the perception and consequences of their own behavior, it is hard to see how a political system of any kind could survive except by chance in the chaotic world which surrounds it, (Easton, 19656b, Chanan, 1992, Canovan, 1993).

The notion of political community does challenge the notion of legitimate domination in terms of which the state is supposed to function. But it also opens a new justification for Western democracy in terms of power and knowledge. What makes Western democracy unique, one might claim, is not so much its method of government as the relative autonomy of its citizens which has developed in its political community more than anywhere else in the world - in issue networks, policy communities, various modes of direct citizenship involvement and control, etc. (Etzioni-Halevi, 1993). What counts for Western democracy is not simply the development of 'ruling classes', 'interest groups' or 'civil society' but the mutual and reciprocal control of rules and resources by political authorities and citizens in the practices or institutions comprised by their relationships. The only problem is that lay-citizens do not enjoy the same degree of freedom and equality in the political community as do elites and sub-elites.

However, the lack of lay-citizenship control may be filled by instituting a new mode of authority in the political system which operates in terms of the "logic of appropriateness" which guarantees that "rules are followed and roles are fulfilled" (March and Olsen, 1989:161). This precisely communicates that lay-citizens must be afforded the possibility to which are socially constructed, publicly known, anticipated and accepted" (March and Olsen, 1994:4). It indicates that authority is explicitly related to "building community and a sense of common identity within which decisions are made" (ibid:165). But it simultaneously shows that the role of authority is distinct from the role of the citizen, although the actual power of authorities will always depend on the specific relationship between them and non-governmental elites, subelites and lay-citizens in the political community (Wrong, 1979). Authorities may be virtually devoid of political control; they may also be the sources of an independent base of power. However, there will always be those who can assume responsibility for the daily affairs of the system and provide initiative and direction in identifying paradoxes and problems and taking some steps towards their transcendence and resolution (Easton, 1965b:216).

AUTHORITY AS POWER-KNOWLEDGE

The liberal view of democracy as a method of government may have 'won' the struggle against Communism. But it can still lose the battle for further democratization. At least, it seems necessary to complement the protective model of democracy with a more communitarian one, conceiving politics and policy as a mode of being which is at the origin of individual and social rights. Liberal democracy is not the end of all democratization. There is no a priori reason to assume that "[p]ower makes men mad, and those who govern are blind; only those who keep their distance from power, who are in no way implicated in tyranny, shut up in their Cartesian poêle, their room, their meditations, only they can discover the truth" (Foucault, 1980: 51). Yet, liberalism makes this presumption when opposing political power to individual and social freedom. The less politics the more democracy, its general argument runs. In the democommunitarian view above, in contrast, there could be no democracy, unless agents possessed the political capacity to do things otherwise.

Human beings must have a transformative capacity, that is, power, if they are to make their imagination symbolic and real (Giddens, 1979, 1981). Whether this capacity is used to sustain hierarchy or to generate popular control may be more a matter of institutional design than of human character (Brunsson and Olsen, 1993). Authority relations involve such transformative capacity. The recognition and acceptance of its use in society may flow from a number of sources - prudential, instrumental, strategic, communicative, moral, customary, or purely from fear of consequences. Political authority does not imply the existence of a division of quences. Political authority does not imply the existence of a division of interest in society, as Hobbes would have it. It simply reveals that a political system exists in society which possesses the capacity to articulate and implement collectively binding decisions. No more; no less.

Authority is not to be thought of as external to the day-to-day life of 'ordinary' citizens but as manifesting their abilities to 'go on' in a practical, taken-for-granted way in the political system, routinely 'rationalizing' what they do. As a citizen, it is perfectly feasible to recognize and accept the binding power of authority and at the same time self-consciously and actively resist those who employ it to accumulate power solely for the purpose of accumulation 'itself' 'as such' (Wilden, 1972). This is merely a matter of distinguishing the power-knowledge of authority from those who put it to use in order to appropriate command over others. 'All' authority does is to point up the difference between those who are senders and receivers of its communicated messages. As David Easton wrote in 1965:

> Unfortunately, in political research we have no convenient term for distinguishing the authorities from all other members in the [political] system. Marx's ruling class as against the ruled, Pareto's elite, Mosca's political class and Michels' oligarchy versus masses are transpararently not satisfactory for this purpose. They classify members of a system according to the power they hold whereas here we wish to point up the difference between those who are occupants of authority roles as against the occupants of all other roles. But however we might classify members in a systematic structural analysis of political systems, this much can be said...: the authorities need not be co-extensive with the politically relevant members (1965b:214-215).

Political authorities are not only capable of dominating local political community action but also of supporting and facilitating it (Chanan, 1992). There are many reasons for assuming that the struggles of citizens for democratizing their daily life will depend for its effectiveness on the capacity of authorities to respond in a way they consider appropriate to cope with the conditions they perceive (cf. Easton, 1965b: 468). Authority cannot be restricted to the uniform authority of the state. It does not imply that demands must always be accumulated or passed upwards in the political system, so that no changes can be introduced internally or locally (Bang, 1987 a+b). Authority finds its opposite in patternlessness, not in anarchy or disorder. It comprises continuities of political form which exist in times of comprehensive social transitions as well as in times of perfect social harmony.

Authority is a type of power-knowledge which occurs (1) when a citizen receives an explicit message from an authority; (2) when s/he then accepts it as the basis for decision and action; and (3) when the grounds for doing so spring from the practical recognition that messages received in this way must or ought to be obeyed without evaluating the merits of the proposed conduct in the light of one's own discursive standards of judgments (cf. Easton, 1955: 28-29). Violence, manipulation and persuasion do, of course appear regularly in the empirical political system. But distinct from authority these other kinds of power-knowledge do not appear to be necessary for making and implementing collectively binding decisions. It is likely that a political system could operate on the basis of authority alone. But it is highly unlikely that it could keep itself socially entirely on the basis of violence, manipulation or persuasion. Even used in one or the other combination they would be much too costly in their implementation to be feasible.

In the authority relationship, B is expected to acknowledge a message from A as binding and to accept and carry it out in practice without further reflections on its merits. That is to say, B must know what A's intentions are in order to be able to act upon A's message in practice. But B need not come up with a principled discursive account of their merits for his or her doings or refrainings. Unlike manipulation and persuasion, authority does not function through B's unconscious or conscious wants (motivations) but via B's practical consciousness of 'what has to be done'. This may in a certain sense be unconscious, because it consists in forms of knowledge not immediately available to B's discursive awareness (Giddens, 1987:63).

Table 4. Types of power-knowledge

	power		
	manipulation	persuasion	authority
knowledge			
discursive	A	A + B	A
practical			B

Political authority, as Table 4 indicates, is a condition of the emergence of political form. No political authorities would ever be able to find the time or energy to beat, fool or convince all citizens all of the time in order to get them to do what they would otherwise not have done. Recognition and acceptance of political authority seem inescapable. As opposed to violence, authority does not make use of direct physical or psychic force. As distinct from manipulation it presupposes capability and knowlegeability on the part of both the sender and the receiver of its communicated message. And in contrast to persuasion it does not bind the sender to engage in a moral dialogue about its worthiness; nor does it require that the receiver evaluates the communicated message in the light of hers or his own standards of judgment (Easton, 1955, 1958).

When modern discourse theorists talk about "the *shamming* of communicative relations in bureaucratically dessicated, coercively harmonised domains of pseudo-democratic will-formation" (Habermas, 1982: 282), they are consequently not referring to the power-knowledge of authority at all. They are simply describing the built-in tension in modern society between instrumental reason and communicative ethics, discursive manipulation and discursive persuasion. As shown in Table 4, discursive manipulation relies on a one-way relationship between A and B where B is unconscious of what A is consciously attempting to get her or him to do. Discursive persuasion, in contrast, presupposes that both A and B can account for what they are doing in order for them to reach a normative agreement in dialogue. Authority, however, cannot function through systematically distorted communication. A must openly communicate the order to B, since B would not be able to obey it, unless s/he understood what had to be done. But B does not have to engage in a dialogue with A to demonstrate the subtle and dazzlingly complex forms of practical knowledge embedded in, and constitutive of, the orders s/he carries out.

The day-to-day acceptance and recognition of authority by lay-citizens may be neither 'planned', as a goal-directed form of individual action (discursive manipulation) nor 'linguistically coordinated', as a consensual form of social interaction (discursive persuasion). But this does not imply that citizens are governed by the irrational parts of their personality (Parsons, 1951). Citizenship always implicates practical knowledgeability, as it occurs in the application and continued reformulation of 'what everyone knows'. This consists of mutual *knowledge* of convention which actors must possess in common in order to make sense of what both they and other actors *do* as participants in a political community. There is a vast conceptual arena here for the reintroduction of the lay-citizen as a capable and knowledgeable subject, whose activities are geared to the continuities that exist in the political system and whose knowledgeability is expressed in the decision-making processes that manifest them (Giddens, 1987:64).

Authority is a continuous threat *and* possibility. It is contingent on domination *and* freedom, and it can be sanctioned by the methods of 'the stick' (the threat of violence), 'the deal' (the promise of rewards) or 'the kiss' (the inspiration of trust and loyalty, Hague et al 1992:10). Authority makes it possible to discipline without weapons, to coerce without systematically suspending B's judgment, and to intervene in the name of the common will, without continuously demanding citizens to engage in a costly dialogue about its justifiability. It is a constituent of, rather than a barrier to, knowledge, showing that in politics "there is no power relation without the correlative constitution of a field of knowledge, nor any knowledge that does not presuppose and constitute at the same time power relations. These 'power-knowledge relations' are to be analysed, therefore, not on the basis of a subject of knowledge who is or is not free in relation to the power system. On the contrary, the subject who knows, the objects to be known and the modalities of knowledge must be regarded so many effects of these fundamental implications of power-knowledge and their historical transformations (Foucault, 1979, 1975: 27-28).

Hence, hierarchy *is* the problem rather than the solution in Western democracy. It creates the vertical and horizontal extension of imbalanced disciplinary and regulatory power to more and more areas of private and social life. This problem does not disappear by rolling the state back as a 'small' but 'strong' *Leviathan* contracting politicians and bureaucrats to act in the interest of private citizens as 'principals' (cf Lane, 1992:7). Why should citizens freely surrender their power and judgement to the state? It really makes no sense at all (Bang and Jacobsen, 1994). Such freedom would not only be a contradiction in terms. It would also conceal the presence of hierarchical discipline and regulation outside "that *sphere* [in which] the exercise of authority and, more broadly, imperative co-ordination, consists precisely in administration" (Weber, 1947:333, my italic). The effectiveness of discipline and regulation, as a mode of political domination, stems precisely from their exercise and routine and silent operations in expert systems which are not legally circumscribed to pay any attention to lay-actors' capabilities and knowledgeabilities as citizens.

Local political analysis should dissociate itself from the famous Parsonian dictum that "[i]f power has inputs of support without [economic] capacity, it will not be effective. If, on the other hand, it has inputs of capacity without [social] support, it will not be authoritative" (Alexander, 1984:90). Inputs of economic facilities have, of course, been vital to expanding the *administrative* capabilities of the state. But they do not in themselves explain developments in the *political* abilities of leaders to decide on the appropriateness of a given course of action. In the same fashion social support of the rules and procedures by reference to which "adherence" to a pattern of values is translated into implementing action has proved important to legitimating political action (Alexander, 1984, Parsons, 1951). But it cannot replace the *politically generated* support of difference ('trust') by which implementation is tied to the expansion of agents' life chances and capabilities to govern themselves in the political community (March and Olsen, 1989).

However strong the norms of rational efficiency in handling politics and policy may be, "they need to be reinforced by specific routines to detect and break away from outmoded prejudices and habits" (Easton, 1965b:461). On the other side, "no system is so homogeneous or consensusal that all the members necessarily conform to precisely the same norms" (ibid:105). The timeliness and effectiveness of political decisions and actions are not primarily the result of individual calculations of utilities but of "the level of education, experience and good judgment to be found in the specific occupants of authority roles at any moment in history" (ibid:-451). In the same fashion, the 'internalization' of common values and norms cannot replace the formation of decisions in the public parlor of the citizen. Citizens in the political community who are responding to what the authorities have done are neither cultural 'dopes' nor economic 'maximizers'. They "are part or all of the politically relevant members in a system, those who can and do participate in the political processes" (ibid, 1965b:401). Thus, "[w]e should not cede tradition to the conservatives!" (Giddens, 1979:7). Democratic traditions and routines in support of the exercise of political difference seem as essential to societal development as a market economy or civic values and norms. The latter can in no way replace the institutionalization of democratic practices for exercising appropriate leadership and for routinely taking active co-reponsibility as citizens for the way values are allocated authoritatively for society.

FINAL REMARK

In this study, I have attempted to dissociate local political analysis from the 'Great Narratives' of liberalism and socialism as being unable to capture the nature of political authority as a type of power-knowledge intrinsic to keeping a political system going in society in one form or the other (cf. Easton, 1965a+b). I have sought to replace the old methodologies of individualism and collectivism by a new institutionalism, oriented towards the development of new political and democratic models of governance at the meso-level (Olsen, 1992). I have also tried to break local political studies free of the juridical and negative representation of political power-knowledge as power/law or power/sovereignty for the sake of capturing the interdependence between political authorities and citizens in the political community (Foucault, 1990, Easton, 1990).

My argument has been that neither the 'exit' strategy of the capitalist market place nor the 'voice' strategy of the civic culture can fill the whole in lay-citizenship control in the Scandinavian welfare state. Their logics of consequentiality and consensus do not cover the logic of appropriateness and sound judgment inherent the making and the implementing of a risky, authoritative decision. Rather they indirectly help in sustaining the political dominion of lay-citizens by various expert institutions. They obstruct popular involvement and control in presupposing that citizens have (or can have) immediate access to the information required to make a rational choice and to reach a normative agreement in dialogue. They reinforce established asymmetries of power and knowledge in the political community by demanding all agents to become 'experts' who can always express their reasons for what they do in verbal or discursive form. They prevent us from seeing everyday skill and knowledgability as standing in a dialectical connection with the colonizing effects of expert systems, continuously influencing and restructuring the very impact of such systems in day-to-day existence in the political community (Giddens, 1991: 138).

All in all, my analysis has called attention to the need for a new comparative political approach to local politics and policy which explicitly recognizes (a) the *double-hermeneutics* of political authority as a medium for relating the discursive articulation of politics and policy to the tacit or taken-for-granted qualities of day-to-day political life in the political community (Giddens, 1976); (b) the *duality of structure* implicated in the structuring of the democratic regime as both medium and outcome of political practices of leadership and citizenship (Giddens, 1979); and (c) *the dialectic of control* between authorities, elites, sub-elites and lay-citizens which makes it possible for even the 'weakest' of political agents to turn some resources back against 'the strong' (Giddens, 1981).

I have argued, a democratically oriented approach to the locality, must operate on the prejudice that novelties appear in Praxis and are mostly *done* knowledgeably as embedded in, and constitutive of, the activities that citizens carry out (Giddens, 1979). The point of doing political research, from a practical angle, cannot solely be to enable politicians and policy-makers to better understand the social world and thereby to steer it in a more reliable fashion than would otherwise be the case. As 'parts' of the citizenry, and, therefore, of our own problems, we have to acknowledge that *in a communitarian democracy* the most effective forms of connection between political research, politics and policy-making are forged through extended communication between political authorities, researchers and those affected by whatever issues are under consideration (Giddens, 1987). Of course, in actual affairs, rulers, formal or effective, will often do their utmost to conceal the dialectic of control inherent to authority (Flyvbjerg, 1991). But the extent to which they succeed crucially depends on the citizenry's will to decision and action. In practice, it is hard to see, how there can be a 'true' democracy, if political 'Herrschaft' is either 'natural' or 'ordained'. A democrat simply has to believe in the possibility that the citizenry really can become able to govern themselves. Otherwise how can s/he *be* a democrat in the first place?

REFERENCES

Alexander, Jeffrey, C. (1984) *The Modern Reconstruction of Classical Thought: Talcott Parsons*. London: Routledge & Kegan Paul.

Andersen, J., A.-D. Christensen, Langberg. K, Siim, B. and Torpe, L. (1993) *Medborgerskab: Demokrati og politisk deltagelse* (Citizenship: Democracy and Political Participation). Herning: Systime.

Andersen, Bent Rold (1991) *Velfærdsstaten i Danmark og Europa*. Copenhagen: Fremad.

Baldersheim, Harald & Illner, Michael (1994) "Local Democracy in East-Central Europe". Berlin: IPSA World Congress.

Bang, Henrik P. (1987) 'Politics as Praxis', *Statvetenskaplig Tidskrift*, no 1: 1-20).

Bang, Henrik P. (1987) 'The Reawakening of a Slumbering Tradition: A Reply to Margareta Bertilsson', *Statvetenskaplig Tidskrift*, no 4: 303-312.

Bang, Henrik P. and Dyrberg, Torben Bech (1993) 'Hegemony and Democracy', *Department of Economics, Politics and Public administration, Aalborg University*.

Bang, Henrik P. and Jacobsen, Uffe (1994) 'The Tradition of Democratic Socialism: A Critique of Liberal Realism', *Statvetenskaplig Tidskrift*, no 1.

Barry, Brian and Hardin, Russell (1982) *Rational Man and Irrational Society*. Beverley Hills: Sage Publications.

Bellamy, Richard (ed.) (1993) *Theories and Concepts of Politics*. Manchester: Manchester University Press.

Bernstein, Jay in Osborne (ed.) (1991) "Right, Revolution and Community".

Blackburn, Robert (ed) (1991) *After the Fall.* London:Verso.

Blondel, Jean (1990) *Comparative Government.* Hemel Hempstead Hertfordshire: Simon and Schuster.

Bourdieu, Pierre (1990) *The Logic of Practice.* Cambridge: Polity Press.

Brunsson, Niels and Olsen, Johan P. (1993) *Reforming Organizations.* London: Routledge.

Buchanan, J.M. (1987) *The Constitution of Economic Policy.* Stockholm: Nobel Foundation.

Burchell, Graham/Gordon, Colin/Miller, Peter (eds.) (1991) *The Foucault Effect: Studies in Governmentality.* London: Wheatsheaf.

Canovan, Margaret (1993) *Hannah Arendt.* Cambridge: Cambridge University Press.

Chanan, Gabriel (1992) *Out of the Shadows. Local Community Action and the European Community.* Luxembourg Office for Offical Publications of the Political Community.

Christiansen, Peter Munk (1993) *Det frie marked - den forhandlede økonomi.* Copenhagen: Jursist og økonomforbundets forlag.

Clark, Paul Barry (ed.) (1994) *Citizenship.* London: Pluto Press.

Crook, Stephen/Pakulski, Jan/Waters, Malcolm (1992) *Postmodernization: Change in Advanced Society.* London: Sage.

Dalton, Russell J. and Kuechler, Manfred (eds), (1990) *Challenging the Political Order.* Cambridge: Polity Press.

Damgaard, Erik (ed.) (1986) *Dansk demokrati under forandring.* Copenhagen: Schultz.

Dente, Bruno and Kjellberg, Francesco (eds.) (1988) *The Dynamics of Institutional Change.* Londo: Sage.

Donzelot, Jacques (1991) "The Mobilization of Society" in Burchell, Graham/Gordon, Colin/Miller, Peter (eds)

Easton, David (1953) *The Political System.* N.Y.: Alfred A. Knopf.

Easton, David (1955), 'A Theoretical Approach to Authority'. *Office of Naval Research,* Report 17:1-59.

Easton, David (1957) 'An Approach to the Analysis of Political Systems'. *World Politics,* no. 9.

Easton, David (1958) 'The Perception of Authority and Political Change' in Friedrich, C.J. (ed.):170-196.

Easton, David (1965a) *A Framework for Political Analysis.* Englewood Cliffs: N.J. Prentice Hall.

Easton, David (1965b) *A Systems Analysis of Political Life* N.Y.: Wiley and Son.

Easton, David (1990) *The Analysis of Political Structure* N.Y.: Routledge, Chapmanm and Hall, Inc.

Elklit, Joergen and Tonsgaard, Ole (eds) (1985): *Valg og Vælgeradfærd.* Aarhus: Politica.

Esping-Andersen, Gösta (1985) *Politics against Markets: The Social Democratic Road to Power.* Princeton, N.J.: Princeton University Press.

Esping-Andersen, Gösta (1991) *The Three Worlds of Welfare Capitalism.* Cambridge: Polity Press.

Etzioni- Halevy, Eva (1993) *The Elite Connection.* Cambridge: Polity Press.

Eyerman, Ron and Jamison, Andrew (1991) *Social Movements: A Cognitive Approach.* Cambridge: Polity Press.

Flyvbjerg, Bent (1991) *Rationalitet og magt I + II.* København: Akademisk Forlag.

Fonsmark, Henning (1990) *Historien om den danske utopi.* Copenhagen: Gyldendal.

Foucault, Michel (1980), (1972) *Power/Knowledge,* edited by Colin Gordon, N.Y.: Pantheon Books.

Foucault, Michel (1979), (1975) *The Birth of the Prison.* N.Y.: Vintage Books.

Foucault, Michel (1990), (1976) *The History of Sexuality: An Introduction, Volume I.* N.Y.: Vintage Books.

Friedrich, C.J. (ed.) (1958) *Authority.* Harvard University Press.

Gadamer, Hans-Georg (1981). *Reason in the Age of Science.* Cambridge, Mass: The Mit Press.

Giddens, Anthony (1976) *New Rules of Sociological Method.* London: Hutchinson.

Giddens, Anthony (1979) *Central Problems in Social Theory.* London: MacMillan.

Giddens, Anthony (1981): *A Contemporary Critique of Historical Materialism.* London: MacMillan.

Giddens, Anthony (1985) *The Nation State and Violence.* Cambridge: Polity Press.

Giddens, Anthony (ed:) (1986) *Durkheim on Politics and the State.* Cambridge: Polity Press:1986.

Giddens, Anthony (1987) *Social Theory and Modern Sociology.* Cambridge: Polity Press.

Giddens, Anthony (1990) *The Consequences of Modernity.* Cambridge: Polity Press.

Giddens, Anthony (1991) *Modernity and Self-Identity.* Cambridge: Polity Press.

Goetz, Edward G. and Clarke, Susan E. (eds.) (1993) *The New Localism.* London: Sage

Gray, Clive (1994) *Government Beyond the Centre*. London: The MacMillan Press

Gundelach, Peter and Siune, Karen (eds.) 1992: *From Voters to Participants*. Aarhus: Politica

Gunnell, John G. (1986) *Between Philosophy and Politics: The Alienation of Political Theory*. Amherst: University of Massachusetts Press.

Gunnell, John G. (1993) *The Descent of Political Theory: The Genealogy of an American Vocation*. Chicago: University of Chicago Press.

Habermas, Jürgen (1982) "A Reply to My Critics" in Thompson and Held.

Habermas, Jürgen (1987) *The Philosophical Discourse of Modernity*. Cambridge: Polity Press.

Held, David (1987) *Models of Democracy*. Cambridge: Polity Press.

Held, David (1989) *Political Theory and the Modern State*. Cambridge: Polity Press.

Held, David & Thompson, John B. (eds.), (1989) *Social Theory of Modern Societies: Anthony Giddens and His Critics*. Cambridge: Polity Press.

Held, David (ed.) (1991) *Political Theory Today*. Cambridge: Polity Press.

Held, David (ed.) 1993 *Prospects for Democracy*. Oxford: Polity Press.

Heywood, Andrew (1994) *Political Ideas and Concepts*. London: The MacMillan Press.

Inglehart, Ronald (1990) "Values, Ideology, and Cognitive Mobilization in New Social Movements" in Dalton and Kuechler.

Isaac, Jeffrey C. (1987) *Power and Marxist Theory: A Realistic View*. N.Y.: Cornell University Press.

Joergensen, Henning et al (1992) *Medlemmer og meninger*. Ålborg: LO & Carma.

Joergensen, Henning et al (1993a) *Sikke nogen typer...* Ålborg: LO & Carma.

Joergensen, Henning et al (1993b) *Fællesskab og forskelle*. Ålborg: LO & Carma.

Keane, John (1988) *Democracy and Civil Society*. London: Verso.

Keane, John (1991) *The Media and Democracy*. Cambridge: Polity Press.

Knorr-Cetina, Karin D. and Cicourel, Aaron V. (1981) *Advances in social theory and methodology: Toward an integration of micro- and macro-sociologies*. Boston: Routledge and Kegan Paul.

Kooiman, Jan (ed.) (1993) *Modern Governance*. London: Sage.

Lane, Jan-Erik (1993) *The Public Sector*. London: Sage.

Lane, Jan-Erik and Ersson, Swante (1994) *Politics and Society in Western Europe*. London: Sage

Laclau, Ernesto (1990) *New Reflections on the Revolution of Our Time*. London: Verso.

Larsen, Bøje and Häuser, Ivan (eds.) (1991) *Decentralisering i praksis*. Charlottenlund: Forlaget sporskiftet.
March, James G & Olsen, Johan P. (1989) *Rediscovering Institutions*. N.Y.: The Free Press.
March, James G, & Olsen, Johan P. (1994) "Institutional Perspectives on Political Institutions". Congress paper. Berlin: Ipsa Conference.
McLennan, David and Sayers, Sean (eds.) (1991) *Socialism and Democracy*. London: Macmillan.
Mouffe, Chantal (ed.) (1992) *Dimensions of Radical Democracy*. London: Verso
Mouffe, Chantal (1993) *The Return of the Political*. London: Verso
Mulhall, Stephen and Swift, Adam (1992) *Liberals and Communitarians*. Oxford: Blackwell Publishers.
Olsen, Johan P. (1991a) 'Rethinking and Reforming the Public Sector', *LOS-senter Notat*, no. 33:1-28.
Olsen, Johan P. (ed.) (1991b) *Svensk Demokrati i Förandring*. Stockholm. Carlssons.
Olsen, Johan P. (1992) 'Analyzing Institutional Dynamics', *LOS-senter Notat*, no. 14: 1-38.
Osborne, Peter (ed.) (1991) *Socialism and the Limits of Liberalism*. London: Verso.
Page, Edward C. (1991) *Localism and Centralism in Europe*. Oxford University Press.
Parsons, Talcott (1951) *The Social System*, N.Y.: The Free Press.
Pedersen, Ove K., Andersen, Niels Aa., Kjær, Peter and Elberg, John (1992) *Privatpolitik* (private politics). Copenhagen: Samfundslitteratur.
Pedersen, Ove K. et al (1994) *Demokratiets lette tilstand*. Copenhagen: Spektrum.
Reed, Michael (1985) *Organizational Analysis*. London: Tavistock Publications.
Rousseau, Jean-Jacques (1978, 1762) *The Social Contract*, N.Y. St Martin's Press.
Schwartz, Nancy L. (1988) *The Blue Guitar*. The University of Chicago Press.
Selle, Per (1990) "Ideen om likskab og tryggleik i nordisk sosialdemokrati". *Los-senter Notat* 90/33
Squires, Peter (1990) *Anti-Social Policy: Welfare, Ideology and the Disciplinary State*. N.Y. Harvester Wheatsheaf.
Stenbergen, Bart van (ed.) 1994 *The Condition of Citizenship*. London: Sage
Thomsen, Niels (1992) *Vi og vore politikere*. Copenhagen: Spektrum.

Thompson, John B. and Held David (eds.) (1982) *Habermas: Critical Debates*. London: Macmillan.

Viotti, Paul R. and Kauppi, Mark V. (1990, 1987) *International Relations Theory: Realism, Pluralism, Globalism*. N.Y.: MacMillan.

Waever, Ole, Buzan, Barry, Kelstrup, Morten and Lemaitre, Pierre (1993) *Identity, Migration and the New Security Agenda in Europe*, London: Pinter Publishers.

Weber, Max (1947) *The Theory of Economic and Social Organization*. N.Y.: The Free Press.

Wilden, Anthony (1972) *System and Structure*. London: Tavistock Publications.

Wrong, Dennis H. (1979) *POWER: Its Forms, Bases and Uses*. Oxford: Basil Blackwell.

Chapter 4

Local Institutional Change in Sweden - A Case Study*

Stig Montin and Gunnar Persson

1. Introduction

As in the other Nordic countries, there have been three waves of decentralization since the late 1960s in Sweden (cf. Bogason 1993). First, there was a transference of functions and responsibilities from the central government to the amalgamated communes. Second, some twenty communes introduced neighbourhood councils, thus devolving parts of their functions one step further down. The third wave started in the late 1980s when communal functions were increasingly privatised or transferred to interest organizations and other actors in the civil society.

At the local government level the post-modern history can also be described by focusing on how dominating reform actors have formulated the problems since the mid 1970s. The first problem was defined as: *too little democracy and too weak political control.* In the legislative material pertaining to the 1977 Local Government Act, and in the reports of the subsequently established Local Government Democracy Committee, there was discussion of various ways of giving people greater influence on communal policy-making. From this resulted an Act granting the communes the right to create local decision-making boards (i.e. neigh-

* This article is based on research sponsored by the Swedish Council for Building Research (proj. no. 930163-9) and the Swedish Association of Local Authorities.

bourhood councils). There was also discussion of the role in local government of elected representatives and the political parties. Research reports and the findings of investigations drew attention to the increased influence of local government politicians. As the local authorities increased their administrative competence and the administrations became professionalised, there was increasing concern about the risk of "bureaucratic rule". Various measures were taken to strengthen the role of the elected representatives in their relationship to the administration.

In the mid-80s the problem formulation became the following: *too little efficiency and too much political control of details*. Whereas before it had been a question of democratisation through decentralisation, it now became more a question of increasing efficiency through decentralisation and delegation. More efficient political and administrative leadership was expected in the communes. Management ideology had a great impact. Whereas for instance a day nursery supervisor had previously had pedagogic responsibility but had otherwise been just one more member of the team, he or she was now to be the head, with economic and staff responsibility. Executive development courses became common and management by objectives the catchword. It was expected to lead to more effective activity, flexibility, better political control and increased civil commitment. The politicians would not occupy themselves with details but would indicate clear goals and afterwards do follow-ups and evaluations.

After a few years - at the end of the 1980s - the problem was reformulated as being the following: *too much politics and too much public sector*. The so-called orderer-performer model (*beställar-utförarmodellen*) was introduced, its purpose being to draw a sharp line between politics and production (Montin 1992). The idea was that the politicians should be considerably less involved in the doings of the service production side, i.e. production should be freed from politics. From this to privatisation was not a big step, and indeed certain local authorities tried to move towards a reduction of their responsibility for certain operations. It was customer choice that should decide what was to be produced, not political decisions.

Finally, more recently the discussion concerning the communes has revolved around the following problem formulation: *too weak civil society*. Inspired by the soft revolution in Central and East Europe, several ideology producers argue that real democracy could only be built in the so-called third sector. Other, more pragmatic actors are trying to find new ways to mobilize resources among voluntary organisations as a complement to the public sector. The idea of user participation and responsibility for communal services can be seen as an ingredient in the efforts to strengthen the role of civil society.

The aim of this article is to take a closer look at the development in one particular commune in Sweden which can be a good illustration of the above-mentioned change of problem formulations. In Örebro many efforts have been made to change the political and administrative structure in order to enhance effectiveness, efficiency, local democracy and legitimacy. The commune was during the 1950s and 1960s well known for its high social ambitions, especially within housing policy. A couple of housing estates were seen as exemplary in the professional literature because of the combination of good residential environment and an ideology of neighbourhood (Elander, Strömberg, Danermark & Söderfeldt 1991, p. 189). The commune also became famous for its early decentralization reforms. Örebro was one of the first three communes which introduced neighbourhood councils and was during the first years of the 1980s a meeting place for all other interested communes from all over the country. In the early 1990s the third wave of decentralization became institutionalized in Örebro by the introduction of user boards. The idea of user boards, e.g. for schools and day care centres, was not a new thing, but the strategy became more serious than before.

First we discuss some of the most important general problems the communes in Sweden are currently facing. After that we describe the development from political and administrative decentralization to market-orientation and the rediscovered interest in user participation and user democracy in Sweden in general and in Örebro in particular. We define three strategies concerning the relation between citizens and local political and bureucratic authorities, and illustrate the theoretical discussion by presenting some empirical findings from a case study of user boards in Örebro.

2. Decentralised Welfare State with Financial Problems

Sweden is in large part a decentralised welfare state. The expansion of the welfare state services in Sweden can be reconceptualized as municipal welfare expansion. Local authorities (communes and county councils) have more or less since the 1950s been considered as the most important institutions when it comes to implementing social and educational policies. This does not mean that the autonomy of local government during this entire period has been as strong as it is now. Several decentralisation reforms have made the communes more autonomous.

From the 1950s two boundary reforms have been implemented (1952 and 1974). The number of communes was reduced from about 2,500 in 1951 to 278 in 1974. During the last ten years a small number of communes have been partitionated into two or more units. The number of communes was 286 in 1992.

The growth of local authorities as welfare institutions can be measured in many terms. At most, during the 1960s and 1970s, the growth of volume was in some years about seven to nine per cent. We can compare communal with state consumption. With an index of 100 in 1970 the state consumption rose to about 120 in 1980 and to 130 in 1991 (and has declined after that) and the communal consumption rose to 145 in 1980 and 180 in 1991. 1992 was the first year in modern times when the communal volume declined. Today, commune and county council expenditure accounts for about 25 per cent of the Gross Domestic Product.

Thus during the period of great expansion the communes became strong institutions with substantial financial, legal, political and professional resources (Elander & Montin 1990). In a comparative perspective communes in Sweden belong to a North European group along with Denmark, Norway and Finland. Compared with other countries, such as the UK, France and the US, this category of local government enjoys both a strong constitutional status and relatively high degrees of policy-making autonomy and financial independence (Hesse 1991).

However, the period of expansion seems to be over. After several years of relatively continuous inflow of resources the economic situation in the communes started to change at the beginning of the 1980s. No dramatic reduction of the services was made during the decade but in the early 1990s we can witness that several communes are making substantial cuts in the social and other services. The reduction is forced by both structural-economic and ideological-political considerations. Since the early 1990s, the financial problems of the Swedish communes (and county councils, which are not discussed here) have been exacerbated by sharply rising unemployment. The tax base is growing slowly or not at all, while communal expenditures are increasing (Häggroth et al 1993). The communes are for example responsible for providing social assistance of different kinds in order to ensure that the unemployed still maintain an acceptable living standard.

According to the formerly dominant Social Democratic economic policy the state would be stimulating the economy by for example expanding the public sector in order to enhance the demands for goods and services. However, since the late 1980s this policy has changed and after the election in 1991 when the non-socialist parties came to power one of the main policies was to reduce the role of the public sector and the political institutions at all levels. The communes should have less responsibility for social and other services. "Private" became considered as being more effective than "public", and therefore parts of communal responsibility should be transferred back to "civil society" (individuals, families, co-operatives and private companies). According to this ideology, which started to grow during the 1980s, politics-production separation, competition and consumerism should replace the old ideas of representative democracy. The traditional identity of local government was questioned. Strong forces started to define the communes as service-producing companies rather than political communities (Montin 1993b).

3. PROBLEMS OF LEGITIMACY

Beside the financial problems the communes are facing there are slowly growing problems of legitimacy for the democratic institutions. In Sweden, as well as in other welfare states, there has since the mid 1970s been a growing criticism of the public institutions. They are considered as being (a) rigid and incompetent, with a lack of adaptability; (b) too interventionistic and powerful, and leaving too little space for individual freedom and local autonomy; and (c) dominated by interest-groups. This criticism was supported by the massmedia, political parties (from the left as well as from the right), interest organisations and international organisations (like the OECD) (March & Olsen 1989, p. 97). Often the criticism is contradictory - for whilst the institutions are found to be bureaucratic (undemocratic and inefficient), too resource-consuming and too inclined to embark on extensive encroachment upon civil society, they are at the same time expected to assume social responsibility in providing various types of safety-net for individuals, industry and commerce, and the economy in general.

The discussed problems should, however, in the Swedish context not be confused with a "legitimation crisis" in the sense that large parts of the population became strongly against public institutions. On the contrary, studies of the citizens´ attitudes towards public institutions give the impression that there during the 1980s were no strong demands for "rolling back the state". There was criticism of bureaucracy and intrusion among the citizens, but it was not transformed into a more general loss of public support (Svallfors 1989, pp. 175-176). Hadenius and Nilsson compared the debate among the political elite with the general opinion among the citizens during the 1980s and came to the conclusion that there was criticism of the public sector which was not founded in public opinion (Hadenius & Nilsson 1991, p. 87). Thus there are no strong arguments for claiming that a "legitimation crisis" developed in Sweden from the mid-70s.

The public institutions in general do not suffer from great legitimation problems among the ordinary citizens. But if we focus on the democratic institutions, particularly the political parties and the popular movements (*folkrörelser*) and the political bodies at the central and the local government level, we can observe problems of legitimacy. For at least two decades the traditional party system has been suffering from legitimation problems, such as low recruitment of new members and activists, particulary young people and women, and a reduced capacity to mobilise and activate the population. According to studies of voters' attitudes, between 1968 and 1991 the "distrust" of the political parties increased continuously (Gilljam & Holmberg 1993, p. 170). According to several research results and other investigations, participation in local branches of established parties and contacting local representatives in order to discuss political matters are declining activities. This does not mean that political interest among Swedes has declined; it means that they use other channels than the traditional parties (Strömberg & Westerståhl 1984; Petersson, Westholm & Blomberg 1989, p. 92).

Looking at the local government level there were fewer voters in 1991 than in 1979 who considered it efficient to influence the overall policy-making by voting for communal council members (from 74 per cent to 61 per cent) (*Kommunaktuellt , No 39, 1993*). The trust in personnel and the service organisations is higher than the trust in the local politicians. This fact can be interpreted as indicating that the base of legitimacy of the politicians is more indirect than direct. Furthermore, about 15 per cent of the adult citizens are interested in being elected to a communal committee or the communal council but about 40 per cent are interested in joining different kinds of user boards for specific services, like a school board (*Kommunaktuellt, No 1, 1994*).

The mentioned observations of the recent development concerning the status of the established democratic institutions may be interpreted as indicating that "(c)itizens are all for a democratic regime but mistrust parties and other institutions that should carry and develop it" (Hjern 1992, p. 1). Thus, there are financial problems behind the reforms and experiments concerning organizational matters at local government level from the beginning of the 1980s, but in the long run legitimacy problems seem to be the strongest driving force.

4. Different Strategies

There can be made a considerable list of different responses to and strategies for handling the above problems. We will, however, focus on three strategies that directly concern the relation between the citizens and the local authorities: neighbourhood councils, user democracy and freedom of choice.

4.1. Neighbourhood Councils

One response to the discussed problems of legitimacy, efficiency and democracy is to try to vitalize the political inflowside of the local political system, e.g. tryi to strenghen the role of the local branches of the political parties. This strategy puts the stress on the citizen as political actor. The ideal is - simplified - that if people can be motivated to engage in local politics within the parties it will make the whole commune more vital as a democratic institution. Participation in local matters, it is believed, can enhance the sense of responsibility for handling common affairs at the local level. During the late 1970s and the beginning of the 1980s increased political participation was an explicit goal in the official political language. Thus, the overall objective in introducing neighbourhood councils was to enhance citizen participation.

There were three communes which were ready to establish neighbourhood councils at the same time as a new Act was passed (1979), namely Eskilstuna, Umeå and Örebro. Three years later (1983) twelve communes followed the first ones and five had already chosen to let the neighbourhood councils cover all the area of the commune. The prognosis at this time indicated that a great many of the communes in Sweden would be having neighbourhood councils in a couple of years. In 1993 there were 14 communes with neighbourhood councils which cover the whole geographic area and seven communes with neighbourhood councils wich cover parts of the commune. The total number of neighbourhood council has increased from 50 in 1983 to 140 in 1993.

In 1985 about 65 communes had after investigation choosen not to establish any neighbourhood councils. A year after 47 communes were investigating whether to implement neighbourhood councils but 40 of them decided not to do it. Among the arguments for not introducing neighbourhood councils was that the structure of the commune did not fit, political disagreement among the political parties, a fear of raising costs, resistance among local officials and lack of experience from other communes with neighbourhood councils.

In the early 1980s there were expectations that the neighbourhood council reform would be a very important instrument to enliven local democracy. However, when evaluations suggested that the reform did not come up to the expected level of participation, and when local economic problems became a commonplace, the advocates of the reform redefined the purpose ot it, now emphasizing the efficiency aspects. Thus, today among many local policy-makers the reform is, at best, expected to enhance bureaucratic efficiency, not to increase democratic participation. During the period from 1985 to 1992 eight communes deconstructed their neighbourhood council organisation and started to develop other kinds of organisation models. The official reasons were that the reform increased bureacracy and made decision-making more difficult, that there was weak interest among the citizens, that it entailed the risk of declining professional competence at local level, that it constituted a threat to principles of equality and that it did not give the expected efficiency.

In Örebro, however, the neighbourhood councils were never really questioned. They were established during the Social Demoratic era, but the non-socialist majority had no plans for deconstructing the councils in 1973 to 1976 or in the present period (1991-1994). It is not right to say that the neighbourhood councils were a Social Democratic invention, quite the opposite. There were many opponents among leading members within the party. But in Örebro the leading Social Democrats were strong proponents of the reform. Actually, some of them took an active part in writing the Bill in the central government.

4.2. User Democracy

In the public sector in general, and at the local level in particular, during the 1980s citizens began to be viewed as "users". One of the elements in the new policy of public administration is "users´ democracy". The idea is that parents of children in communal child-care, and also other users, should participate in and exert influence on the production of services. From the point of view of the Ministry of Public Affairs in the mid 1980s all communes were expected to develop this kind of users´ democracy.

The concept of the "user" (*brukare*) is in fact not really a concept of the 1980s. It derives from the Danish concept of the *bruger* and was introduced in the Municipal Democracy Committee at the end of the 1970s. In connection with the work of renewal initiated by the Social Democratic Party it came to serve as a central concept for creating a "complement" to representative democracy (Mellbourn 1986, p. 40).

Before the introduction of this type of user participation there had long been various channels through which special groups could exert influence on public services. At the beginning of the 1970s a number of "councils" were created in most communes: disabled persons´ councils, pensioners´ councils, and adult retraining councils. These can be characterised as corporate bodies inasmuch as they did not represent all the inhabitants of the municipality but only certain interests or client groups (Strömberg & Norell 1982, p. 29).

Despite the fact that user influence was set forth as an important policy during the 1980s, it was not the success that many had hoped it would be. A great number of projects in this direction did start in the municipalities - some of them more successful than others. However, research has indicated that there were several difficulties when it came to developing user influence. There was the problem of defining the users, and that of differentiating between participation and influence; there was conflict between user participation and representative democracy, between user interests and professional interests, and within the "user collective"; and there was the matter of finding practical arrangements for user participation.

To take just one example: in a questionnaire to day nursery supervisors (together with area child care supervisors) and elected representatives in five communes, the respondents were asked to state their position regarding the statement that "There is a lot of talk about parent influence and parent participation but very little happens". 86 per cent of the supervisors and 80 per cent of the elected representatives who answered agreed with the statement (Montin 1993a, p. 138). The results indicate that there is a confusion concerning what form parent participation ought to take, and indicate too that "user participation" in child care in the communes under investigation was apparently more of a slogan than a phenomenon with a specified content.

The restrictions laid on the forms of user influence are clearly associated with the representative democracy, the regulated welfare state, and the roles of the professionals. For example, in connection with the deliberation regarding the new Act at the end of the 1980s it was proposed that user participation should be "regulated" so that it did not come into conflict with other values, in the first place representative democracy (Ministry of Public Administration 1991). However, in spring 1994 an Act was passed in the Riksdag granting the communes to be more free in finding solutions for user democracy.

4.3. Social Democratic Strategy in Örebro

As an effect of a strong wind from the political left all political parties in Sweden considered democratisation of the public sector to be very important during the 1970s. However, the lively debate concerning local democracy resulted in many different strategies. For the Social Democratic Party and the Left Party it was important that national goals like equality should not be threatened by local solutions. The Moderate Party, the Liberal Party and the Centre Party were, on the other hand, arguing for adjusting the democratisation to an old endeavour to reduce the public sector and the role of politicians in favour of private and market-oriented solutions.

After a period dominated by the amalgamation reform and a centralistic model of governance the social democratic leadership in Örebro started at the beginning of the 1970s to introduce a new way of decentralisation and democratisation of communal political decisions. Not least it was considered important to decentralise power *within* the Social Democratic Party, which during many years had been dominated by one strong person. This idea was however not simple to implement. Social democracy in Sweden has a long tradition of belief in the necessity of strong central power as a means for supporting disadvantaged groups in society. The risk of decentralisation was, as it was argued among many party members, that sub-local elites could come in powerful positions and jeopardize the equal distribution of communal services.

The strategy for the decentralist proponents in Örebro was to introduce the idea among party members and the communal staff. User influence got lower priority. There were some experiments of user influence for the tenants in the Communal Housing Company, but it became one of the largest adversities for the social democratic leadership during the 1970s in Örebro. The resistance was strong among their own party members, not least among the party members who also were members of the Tenants´ Interest Organisation. They saw the idea of user democracy as a threat to party influence on housing policy. However, the new decentralist strategy was in many respects successful. Already in 1973 the Communal Council decided to establish four sub-communal boards and eleven years later (1984) the commune of Örebro was ready to leave the period of experiments and took a full step towards creating fifteen neighbourhood councils which covered the whole commune.

The theoretical point of departure was that "every decison should be taken at the lowest possible level". The idea was that a viable democracy cannot be reached if a small number of elected representatives are responsible for all local political issues. Instead the local society should be organised in a way that those affected by the decisions in the neigbourhood should increase their influence. But the influence should be channelled through the local party organisations, not through direct user participation. The representative system with the local branches of the party organisation as sources of recruitment was considered the best and the most legitimate system. By introducing neighbourhood councils the Social Democratic Party hoped for new members and more representatives; many had disappeared after the amalgamation reform.

Ideas of user democracy were indeed formulated, but in practice it was limited to small areas. We have mentioned the idea of tenants´ influence, which met strong resistance. In some cases youth organisations took over the responsibility for youth recreation centres. In some day care centres parents´ participation was introduced, but there were strong protests from the unions that organise the personnel in the day care centres. The concern for equality also slowed down the development towards user democracy in Örebro.

To summarize, decentralisation and democratisation in Örebro during the 1970s and the 1980s was a question of changing the internal political and administrative organisation. Citizen participation outside the traditional political institutions was not considered important. Not until the end of the 1980s were ideas of user participation and democracy really put on the agenda. However, they were challenged by consumer-oriented strategies.

4.4. Freedom of Choice

Since late 1980s we can witness a growing customers´ perspective. Here the stress is put on the freedom of choice (cf. State Commission Report 1993, number 90). The idea is that, for instance, patients should have the right to choose between health centres, and that parents should have the right to choose between schools for their children. The main difference between these two policies can be conceptualized in the words "exit" and "voice" (Hirschman 1970). While freedom of choice is a matter of opportunities to "vote with one´s feet", users´ democracy is a matter of opportunities to exert influence on goals and performance in a given institution. The distinction can also be expressed in terms of "consumer solutions", which emphasize service responsiveness, versus "collectivist solutions", which emphasize service democracy (Hambleton & Hogget 1988:53ff).

We can briefly make a comparative excursion to Denmark. This is relevant because the two countries have similar welfare state systems and because policymakers in Sweden recently have been strongly influenced by different arrangements of user democracy in Denmark. In Denmark - at the time with a non-socialist government - there were at the beginning of the 1980s strong arguments for user participation. Compared to Sweden the interest did not decline, instead several acts were passed in the Danish parliament in favour of user boards and other similar arrangements (Bogason 1993). User participation has never been considered as a threat to representative democracy in Denmark. There is according to some observers a tradition of greater flexibility towards the civil society in Denmark than in Sweden (Arvidsson, Berntson & Dencik 1994, p. 322). On the other hand, consumerism like the idea of defining users as individual consumers of public service does not seem to have been considered as an important alternative in Denmark. There is no such emphasis on freedom of choice and exit possibilities within child care, care of elderly people and primary education in Denmark as in many communes in Sweden.

In 1992 12 communes in Sweden had introduced customer choice for child care, primary school and care of elderly people. According to a recent survey about 60 communes are during 1993-94 introducing different forms of customer choice within the mentioned service areas (Ministry of Public Administration 1994). The market-oriented language is, howerer, widely spread around several communes.

4.5. "NEW START" FOR ÖREBRO

While the social democratic policy got a breakthrough during the 1980s the non-socialist parties in Örebro organised themselves around an alternative policy. After the 1991 election when the right-wing coalition came into power a "new start" for Örebro was proclaimed. The communal service institutions should be set in competition, the cost should be reduced, and new systems of resource allocation were introduced. The policy was to "transfer more decision-making power from communal boards and committees to the individual citizens. Competition and great variety of services will then get more room in the communal planning than before" (Örebro Municipal Council 1992). A quasi-voucher system was introduced within child care and primary school. Private alternatives started to compete with the communal institutions. Drowsily the schools and the day care centres started to produce leaflets with models from the private market. From 1993 all service production was, if possible, to be set in competition. Every committee should make a plan for competition and privatisation. The communal staff were encouraged to develop business concepts in order to start their own business. According to the policy a quasi-voucher system will give the parents freedom of choice. The day care centres and the schools that follow the demands of the parents will then be leading the market. Thus the consumer-oriented system is supposed to be a strong incitement to increased service quality.

In an international perspective the market-oriented strategies in Örebro are not particularly remarkable, but put in a Swedish context the ideas are very different form the traditional views of how to arrange communal social services.

There is a tension between freedom of choice and collective user democracy. The Moderate Party prefer "pure" market solutions but for example the Liberal Party is more ambivalent. One argument against user participation is that it can jeopardize a future privatisation. Strong organisations of parents can slow down the market orientation and even the demand for increased public services. However, it seems that ideas from different ideological standpoints have met around user democracy. The concepts of civil society and voluntary movements seem to have tempted people from different political directions.

5. USER BOARDS IN ÖREBRO - A CASE STUDY

The previous discussion points to the fact that there is a vacant space between the citizens and the local politicial institutions, especially the political parties, and many local reform actors are trying to fill this space with user democracy and consumerist strategies. For about twenty years citizens have less and less used the channels of the political parties. Actions, contacts with the civil servants and other types of political mobilisation have apparently been regarded as more important tools to influence local politics than contacts with the local politicians. Some communes have tried to reorganise their organisation of representatives to facilitate a closer contact between the elected and the voters. Neighbourhood councils and other types of local boards are the most visible monuments of this ambition. After the election in 1991 there has been a wish - especially within the Moderate Party - to develop a democracy without politics. That is, try to reduce the role of politicians in favour of freedom of choice for the communal customers. During the last couple of years there has also been a renewed interest in user boards. These three strategies refer to the citizens as political actors, individual customers and collective users respectively.

As we have seen, all three strategies have been used in Örebro. The neigbourhood councils have been established since 1983 and more recently the ideas of freedom of choice and user democracy have been introduced. Our interest is focused on what will happen with the local political dialogue when user boards are introduced in a specific area in the commune of Örebro. Do, for example, the members of the boards see themselves as involved in local political matters or do they define themselves as non-political? What kind of relations are to be found between the boards and the sub-communal political and administrative apparatus?

5.1 The Creation of New Boards

There are 15 neighbourhood councils in Örebro and one of them is situated in Adolfsberg-Mosjö. About 10,000 people live in the area. Adolsberg-Mosjö is the most wealthy part of Örebro. Here is to be found a very high average-income, very few people from other countries, a lot of people living in their own houses, very few people living in confined quarters and a high level of tertiary education. Not surprisingly, there has always been a strong support for the non-socialis parties in the area. In the 1988 election the Moderate Party got more than 30 per cent of the votes in some areas in Adolfsberg-Mosjö and in the 1991 election the party got nearly 40 per cent in some areas. In Örebro commune as a whole the Moderate Party received 16 per cent of the votes in 1988 and 20 per cent in 1991. The neighbourhood council is, however, not directly elected. The majority is the same as in the communal council. The Social Democratic Party was in majority in the neighbourhood council until after the 1991 election.

During 1993 the neighbourhood council took a step towards more local democracy by introducing four user boards *(förvaltningsråd)*. The background can be traced to earlier work in the neighbourhood council around the possibility of deepening the democracy around some centres in Adolfsberg-Mosjö. Study trips have been made to Denmark which has similar kinds of local boards. The boards have been initiated by the politicians and the civil servants in the neighbourhood council. Therefore the boards are in the first place a project by the neighbourhood council and hawe not essentially grown from a local opinion outside the political parties. The politicians were searching for a board which could be both advisory and with rights to make decisions on issues concerning the local area. The civil servants in the neighbourhood council were given the task to work out a plan for how these boards should be organised. The administrative chief of the neighbourhood council was a very important prime mover. The boards were established with great support of all the political parties.

In the agreements which were set up between the neighbourhood council and the user boards the following goals can be found:

> "The goal of the Advisory Board is to provide for a good service in the area and to act for an advanced local democracy. The field of responsibility is primarily people between 0-16 years. The board should also act for a good environment in Mosås"

Without its being clearly mentioned it is also to be understood that the boards shall work with all issues connected to the neighbourhood council. But here is to be seen a very clear concentration on the group 0-16 years, that is children and youngsters in school, in the child care institutions and in their free-time activities. But there is also an ambition that the boards will act in other issues, for example remitted questions to do with different issues concerning physical planning.

The neighbourhood council sent leaflets to all households in Adolfsberg to inform everyone about the creation of the boards. The politicians demanded that the users on the board should be elected by a general election in the area. At the information meeting there was an election committee appointed which should recruit the users to the boards. Every area had their own meeting. Only a few of the citizens in the area were present. On the whole, people who showed interest also were elected. A parents´ organisation (*Hem och Skola Föreningen*) had appointed the election committee, which in fact meant that the first generation of users in the boards were elected among the active people in the parents´ organisations. The ambition of the neighbourhood council was that the work of the administrative boards should be orientated towards a broad set of social and other local issues. And the representatives should be recruited from different groups. However, the actual solution was considered as the most practical one for the time being. It can be mentioned that a great many of the households consist of two people without children, more or less 45 per cent, which is remarkable.

5.2. Members and Their Motives

The boards consist of one coordinator, 2-4 people elected from the professional group in the neighbourhood council adminstration and 4-6 representatives from the users. The election period is two years. The boards are expected to meet six times a year, in spare time. Additional meetings are to be held with for example the neigbourhood council. The boards must unite around a decision in consensus (the consensus rule). Otherwise the issue is taken to the neighbourhood council. The boards only have access to money budgeted and given by the neighbourhood council. Most of the meetings which have been held up to late 1993 have had the character of information meetings. It has primarily been information from people working in the area with human services, policemen and so on. Also the politicians on the neighbourhood council have been involved at the information meetings.

Interviews with the users indicate that a very important stance for acting is the role as a parent. A lot of people find it important to be committed to the cause of the children, especially of their own children. Many of the users see their activities on the administrative boards as a continuation of their
activities in the parents' organisations, and, perhaps, a better platform to act from than the position they had in the parents' organisation. Other motives mentioned are "the duty" to care about other people in society, the need to learn more about the political system and budget issues, try to influence in special questions. Some expect themselves to represent the citizens in a proper way, and some of the board members believe that the decisions made in the neighbourhood council will be much better after considering opinions among users on the boards.

A common opinion is that the neighbourhood council politicians and the political parties are not representing the people in Adolfsberg in a good way. According to most of the interviewed members of the boards there are too much political tactics and the elected politicians on the neighbourhood council are too close to the administrative system. The interviewed want the neighbourhood council members to be directly elected instead of appointed within the political parties. One reason is that the political majority on the neighbourhood council is the same as on the communal council of Örebro. Another reason is that they want the neighbourhood council members to discuss the questions without party political loyalties to the centre of Örebro.

The interviewed do not think that members of the boards should be politically elected. The expectation is that the boards will make decisions out of "common sense", not "party politics". Most of the interviewed don´t want to be connected to the politicians. Instead they want to be regarded as ordinary citizens using their common sense and general experience as their way of influence and contribution. A very small part of the persons interviewed talk about some kind of political involvement but none of them referred to membership in a political party as a motive to join the boards. Many of the interviewed mention their professional experience in the private sector as valuable in the otherwise official sphere. They think the dominance of the publicly employed people, also among the politicians, is one-sided and that there is a great necessity for people from the private sector to balance the situation. There is also a difference in the attitude to the relation between decision and action. People coming from the public sector appear to be more understanding that things will take time and that it is good to "avoid unnecessary risks". In contrast the privately employed are impatient at what they see as "bureaucracy and paralysis of action". Thus, the user boards make a good illustration of the tendency that people are interested in getting involved in local common affairs but not in activating themselves in a political party.

5.3. Voices of the Periphery or Instruments for the Centre?

Local institutions, like the user boards, can be initiated from above or below, which is important for their coming role. The neighbourhood council, which initiated the user boards, sees itself as a spokesman for the local community. The boards are often viewed as important voices from the local community, which sometimes can be used as power resources against the communal political and administrative centre.

On the other hand the neighbourhood council is a part of the communal political and administrative system. The obvious risk is that every level in the hierarchy will defend their own interests. It may mean that the neighbourhood council tends to see the boards as a good support for providing human services or legitimate cuts in the services rather than being a defence against the abuse of central power. The way of action on the boards contributes to additional forms of decisions and discussions inside the neighbourhood council´s area of responsibility. New kinds of loyalties and opposition bring people together and apart. The strongest motive for becoming involved in the boards seems to be the possibility of contributing to better service for primarily one´s own children or to try to stop or minimize the effects of the social cuts which are anticipated. If these different motives have not yet been obvious and discussed the coming conflicts will probably be fought out without any direct communication between the boards and the neighbourhood council.

The interviewed board members believed that the neighbourhood council wants them to act mostly as political citizens in the sense that they should deal with many questions at the same time. The neighbourhood council has also given the coordinators "orders" to avoid concentration on one or a few issues only as ordinary users. The user boards are supposed to take an active part in some activity continuously. The role is approaching the role of a manager, which includes more than just making decisions. Sometimes it also means practical work in the service delivery which also will give the user both more feeling for and more knowledge about the activity. In this respect many members of the user boards hesitate. However, it seems that the users are more positive to the active role as a manager if they are involved in a *special* cause. For example in one area, the users want to take an active part in the management of the new school. They have shown a similar ambition earlier with regard to the old school by both purchasing and repairing a new annex of the school.

On the whole, the study indicates that there is a strong connection between influence/power on the one hand and responsibility and commitment on the other hand. That is one reason why the users hesitate to "accept too much influence". The interviewed say that there is a risk that the neighbourhood council wants to get more understanding and support for expected cuts in the human services by introducing local user boards. The boards can help them to explain why the cuts are necessary and get more understanding from the citizens in the area. The users are not sure if the boards imply more influence on the neighbourhood council (the voice of the periphery) or constitute an extension of the neighbourhood council (the latter's antennae or tentacles). Therefore the users are taking a cautious position with regard to the possibility of getting more responsibility. They seem, however to be less cautious when the responsibility appears to include positive changes.

In relation to this there are also tendencies of tension between the users and the coordinators on the boards. The civil servant representatives on the boards are employed and salaried by the neighbourhood council. What role will they get and what happens when opposition occurs partly between them and the users, partly between the neighbourhood council and the users on the boards? There is reason to believe that the civil servants will end up in a difficult situation if they have the same opinion as only one of the other parties. Probably it is easier to take a conflict with the users than the employers. The civil servants have not been obliged to "show their true nature" in any large issues yet, but there is to be discerned some irritation from the users about the civil servants´ information policy in budget issues. Some users believe that the civil servants are not giving enough information in order to avoid discussions about questions not yet decided and difficult to handle.

Shortly after the introduction of the user boards tensions arose between the neighbourhood council and the boards. In 1993 the right-wing majority in Örebro approved a programme for competition which implied that nearly all kinds of communal services should be made subject to the rules of the "market". The neighbourhood council in Adolfsberg-Mosjö was first to adopt the programme. The idea is to set about 20 per cent of the services up for competition every year. Thus after five years nearly 100 per cent of the services no longer will be in the direct control of the neighbourhood council. Only some of the neighbourhood council administration will be kept outside the rules of the market, for example the local town hall officials. The neighbourhood council intends to implement this programme very quickly, but has been met with serious political opposition. Several of the user boards strongly rejected the idea of competition, although their opinion was not requested. The critical voices made the Liberal Party representative doubtful and decide to vote against parts of the programme proposal. The result was that the schools became exepted from the competition programme. The tensions illustrated above call for a basic discussion around levels of influence and responsibility.

5.4. Levels of Influence and Responsibility

After the amalgamation reform during the 1970s the average Swedish commune is large compared with other European countries. An important motive behind the reform was to build effective local welfare institutions with a strong tax base. However, relatively strong state control was considered important for making service quality equal. The central-local relation has not been considered as a dualistic relation. It has rather been thought of as an integrative relation (Kjellberg 1988). Looking at the communal level, equality has been a motive for relatively centralistic policy-making. It has been argued that the risks of inequality and social segregation will increase if the decision-making power is decentralized within the commune. In short, this has been a dominating social democratic view of the centralisation-decentralisation relationship. On the other hand, democracy in terms of extensive citizen participation in representative forms, especially in the local party organisations, has been of high value. Clearly there are conflicts between democracy and equality which have been two main points of departure in Swedish politics, especially during the social democratic era.

The tradition of centralism and representative political control is one factor that explains why the user boards have not yet have been given any budget resources of their own. Another important restriction is that the decisions taken by the boards have to be in full consensus. Otherwise the neighbourhood council alone will take the decision.

The way of recruiting members of communal and neighbourhood councils from the political parties still dominates in Sweden. In Örebro all the members of the neighbourhood councils are appointed within the political parties. Recently, however, it has been questioned, whether it is necessary that sub-local politics should be party politics. Partly this is an ideological question. As we have argued before, neo-liberals want to abolish political control at all levels - i.e. try to build a democracy without politics - while the left wing stresses the importance of a representative political control at all levels. The problem can be illustrated by a figure:

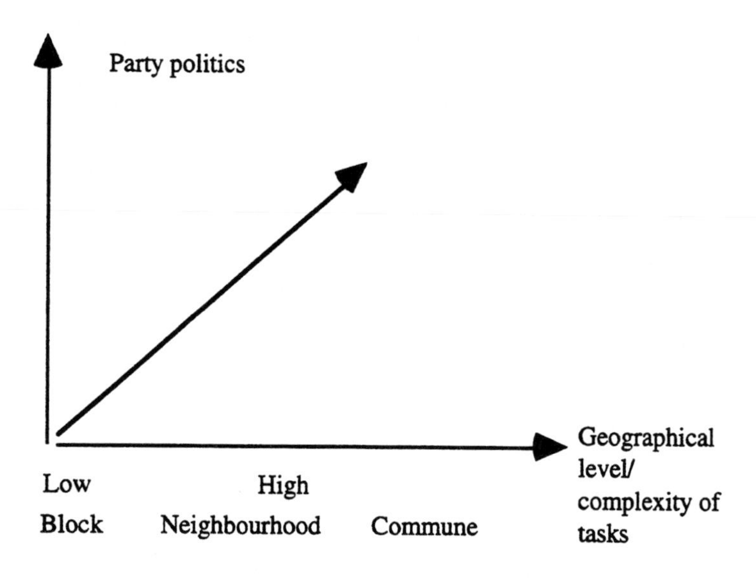

Figure 1. Different types of politics on different levels

People who have accepted the role of party politics will probably see the growing necessity of people elected from the political party system when the the complexity of the tasks is increasing. Here there are differences between people living in the town and people living in the country or villages. The geographical solidarity is stronger in the villages, which also can be seen in Adolfsberg-Mosjö. Often the whole population in the rural areas are united against central powers of any kind.

The top politicians in Örebro resolutely claim that local democracy will be deepened. That every political decision should be taken on the lowest possible level, has been an outspoken goal for a long time.The interviews however indicate that a lot of central decisions are working in the opposite direction. School regulations, purchasing rules, central agreements between the commune and the trade unions, health regulations and security regulations and so forth are examples that have been decided on a higher level for a long time, most of them defined as non-market regulations. Thus, the central-local balance has not yet been solved and the political - non-political border is still a grey zone.

6. Concluding Remarks

Several forces are moving towards a re-learning within the local political and administrative system. There are different responses; one is to make the relationship between citizens and local government more market-oriented, other strategies put stress on local collective decision-making and responsibility. The latter approach is not a new thing, but the interest has grown during the last two or three years, sometimes as a reaction against the market-orientation and individualism.

he briefly presented user boards in Örebro commune are in line with the local collective approach. One of the ideas is that user participation can stimulate local democracy and vitalize the local branches of the political parties, make them more legitimate and integrated in the local community. However, the participants on the board seem to avoid all kinds of party politics in favour of "common sense". There is also a conflict between the boards as an instrument for the neighbourhood council and as a voice for the local community.

On the other hand, the ongoing trend towards consumerism does not seem to have been strongly rooted among the citizens in Adolfsberg-Mosjö. People are more concerned in acting as collective user rather than as individual consumers of communal services. For example, the competition programme did not fall on feertile soil. We might conclude that the study so far indicates that there is an interest in *common* matters but people are not willing to define them as *political* matters.

For further research it is interesting to investigate the relationship between the boards and the local political and administrative system more in detail. For example, we have not yet studied the relation with local professionals and other civil servants in detail. New local institutional arrangements can be looked opon as learning processes. The new arrangements may be formally defined, but after a while when different issues are handled and conflicts and tensions are developing there may be substantial change in relations between different local actors. As we see it, a fruitful research challenge is to study the *dynamics* of local institutional change. This means that stress has to be put on how people actually act when they are meeting different problems and challenges in the local society.

ACKNOWLEDGEMENT

We are greatful to Malcolm Forbes for making the text readable in English.

REFERENCES

Arvidsson, H., L. Berntson and L. Dencik (1991), *Modernisering och välfärd - om stat, individ och civilt samhälle i Sverige.* (Stockholm: City University Press.

Bogason, P. (1993), "The Institutional Development of Danish Human Services", Paper for the Conference on new forms of organization in local government, Oslo, Norway, August 19-21, 1993.

Elander, I., S. Montin (1990), "Decentralisation and Control: central-local government relations in Sweden", *Policy and Politics.* Vol. 18 No. 3, pp 165-180.

Elander, I., T. Strömberg, B. Danermark and B. Söderfeldt (1991), "Locality research and comparative analysis: the case of local housing policy in Sweden". *Environment and Planning* Vol 23, pp 179-196.

Gilljam, M. and S. Holmberg (1993), *Väljarna inför 90-talet.* (Stockholm: Norstedts Juridik).

Hadenius, S. and L. Nilsson (1991), *Ifrågasatt. Offentliga sektorn i debatt och opinion.* (Götebrog: Svensk Informations Mediecenter).

Häggroth, S., K. Kronvall, C. Riberdahl and K. Rudebeck (1993), *Swedish Local Government.* Traditions and Reforms. (Stockholm: Swedish Institute).

Hambleton, R. and P. Hogget (1988), "Beyond Bureaucratic Paternalism", pp 9-28 in Hogget, P. and R. Hambleton (eds), *Decentralisation and Democracy. Localising public services.* (Bristol: School for Advanced Urban Studies).

Hesse, J. H. (ed) (1991), *Local Government and Urban Affairs in International Perspective.* (Baden-Baden: Nomos Verlagsgesellshaft).

Hirschman, A. O. (1970), *Exit, Voice, and Loyalty: Responses to Decline in Firms, Organizations, and States.* (Cambridge: Harvard University Press).

Hjern, B. (1992), "Illegitimate Democracy: A Case for Multiorganizational Policy Analysis", *Policy Currents,* Vol 2, No 1, February 1992.

Kjellberg, F. (1988), "Local government and the Welfare State: Reorganization in Scandinavia", pp 39-69 in Bruno, D. and Kjellberg. F. (eds) *The Dynamics of Institutional change*. Local government Reorganization in Western Democracies. (London: SAGE).

Kommunaktuellt, No 39 1993; No 1 1994.

March, J. G. and J. P. Olsen (1989), *Rediscovering Institutions*. The Organizational Basis of Politics. (New York: The Free Press).

Mellbourn, A. (1986), *Bortom det starka samhället*. (Stockholm: Carlssons).

Ministry of Public Administration (Civildepartementet) (1991), *Vidga brukarinflytandet. En väg till ökad delaktighet och bättre service*. Ds 1991:18. (Stockholm: Allmänna förlaget).

Ministry of Public Administration (Civildepartementet) (1994), *1993 års redovisning av utvecklingen i kommuner och landsting* Ds 1994:61. (Stockholm: Allmänna förlaget).

Montin, S. (1992), "Recent Trends in the Relationship between Politics and Administration in Local Government", pp 31-43 in Batley, R and A Campbell (eds), *The Political Executive: Politics and Management in European Local Government*. (London: Frank Cass).

Montin, S. (1993a), *Svenska kommuner i omvandling*. Novemus rapport 1993:1. (Örebro: Högskolan i Örebro).

Montin, S. (1993b), *Swedish Local Government in Transition*. Örebro Studies 8. (Örebro: Högskolan i Örebro).

Petersson, O., A Westholm and G Blomberg (1989), *Medborgarnas makt*. (Stockholm: Carlssons).

State Commission Reports (SOU) 1993:90.

Strömberg, L. and P O. Norell (1982), *Kommunalförvaltningen*. Ds Kn 1982:8. (Stockholm: Allmänna Förlaget).

Strömberg, L and J Westerståhl (1984), *The New Swedish Communes*. A summary of Local Government Research. (Stockholm: Liber).

Svallfors, S. (1989), *Vem älskar välfärdsstaten?* Attityder, organiserade intressen och svensk välfärdspolitik. Arkiv avhandlingsserie. (Lund: Studentlitteratur).

Örebro Muncipal Council (1992), "New start for Örebro". Declaration from the parties in majority.

Chapter 5

Decentralisation, Privatisation and Representativeness in Local Government

Henry Bäck

Since the mid 1980s local Swedish authorities have gone through extensive changes. These changes have, in part, effected external organisation as part of the central government's explicit aim to increase local authority autonomy. The 'free local authority initiative', the 1992 Local Government Act, the delegation of the school system to the municipalities and a new, more consolidated state grant system are all expressions of this attempt. In practice the effort has not always been consistent. Reforms such as the local authority tax freeze, the cancellation of state grants and the imposition of new burdens due to the transfer of tasks from the central government and county councils have militated against the primary aim (cf. Bäck 1990).

The main organisational changes have, however, affected local authorities' internal organisation. These changes can be described as a transferral of entire operations or central functions in a structure which can be organised along the dimensions of level and sector.

These shifts, which can preliminarily be termed *decentralisation* and *privatisation*, can be expected to have effects upon the way local democracy functions. In this paper the focus is upon how representative politicians are of the electorate. After a discussion of the central concepts - organisational change, representativeness and legitimacy - the results of a research project which examines changing levels of representativeness in Swedish municipalities from the late 1970s is presented.

The results can be summarised: The number of local politicians is dwindling. At the same time the average number of duties per local councillor is rising, and the degree of representativeness in a number of aspects is deteriorating. Dimensions of representativeness investigated include sex, age, class, immigrant status, and area. All these tendencies are more pronounced in those local authorities that have gone furthest toward privatising and marketing their organisation.

Levels and Sectors

Level often refers to a territorial dimension, which has long been neglected in political science (Sharpe 1989). The central level refers to the entire territory. For organisations at the central national level the entire country is their territory of operations. Similarly, local authorities, too, have organisations with responsibility for activities throughout the entire local district territory.

Sector alludes to the conventional distinction made between the public and the private sectors. Both sectors are populated by different types of actors and organisations. Within the public sector is the state and its different components: the political and administrative organs, and the local authorities with their equivalents. Within the private sector appear companies, voluntary organisations and households. A fruitful distinction can be made within this sector between a business and an associational sector, and, perhaps also, between these two sectors, which are dominated by formal organisations, and the informal household sector.

Both dimensions can be combined to yield a matrix in which all of society's organisations have a place. The concept of level is most often explicit in public sector organisations. The government, parliament, central civil service departments and other central state agencies are located at the central level. At the local level are the local authorities, but state agencies are also represented (in the Swedish context e g the police, the tax offices, military units, institutions of higher education, prisons, post offices). A middle or meso level, comprised of regional organs such as county councils etc, is also conventionally distinguished. Within the local authorities local councils, executive boards and committees, as well as central departments are assigned to the central level, while local authority offices and other local organisations, along with individual service providers such as schools, nursery schools and recreational amenities etc. are assigned to the local level.

In comparison, levels within the private sector are less distinct. It is possible though to distinguish structural levels which resemble those found in the public sector within both business concerns and the larger interest organisations. At the local level, households, small businesses, large companies, single production units and local voluntary associations may be included.

POLITICS, HIERARCHY AND MARKETS

A relation exists between sectors and the management or coordination systems. The public sector is dominated by the democratic-bureaucratic chain of command. This has its origin in the needs of the citizens and their wishes for a future state of affairs. The electoral system conveys the values of the electorate to elected representatives who then transform these into political decisions. With the help of a range of management structures, decision makers assume that the professional administrative apparatus and production organisations of the public services will put these decisions into practice. Prior to decisions the coordination principle is politics - a game where power resources and positions are the currency. After the decision the coordination principle is hierarchy - where authority and obedience perform the equivalent functions.

Within the private business sector, the market acts as the coordinating principle between individual actors. Rational, self-interested actors choose between buying and selling alternatives. Different degrees of competitive relations exist in the market ranging from a maximum of concentrated monopoly and monopsony markets to the maximum decentralisation of perfect market competition. In the latter, competition eventually results in the most efficient possible distribution of production resources. Utilities are produced with the minimum possible sacrifice. However, it has long been recognised that there are factors intrinsic to the market which push in the direction of monopolies. Long ago, Adam Smith (1776) noted that as soon as businessmen meet together they conspire against the general public and seek to raise prices.

There are also alternative views of what happens when rational egoists come into contact with one another. The prisoners' dilemma is a well known example in which egoism pushes the system to a state of equilibrium where, unlike in the case of the perfect market, the collective good is not maximised (cf. Zagare 1984, Axelrod 1987).

Just as democracy is characterised by both politics and hierarchy so too is there a dualism within the private sector. Different forms of market interaction coordinate the system of organisations, but within these organisations the principle of hierarchy dominates as in the public sector.

The associational sector displays further similarities to the democratic-bureaucratic chain of command: Members elect decision-making and executive bodies which, at least in larger organisations, have a professional administrative apparatus at their disposal. But there are also important differences between the associational sector and the public sector: Membership is voluntary to a much greater extent within the associational sector where decisions can not be forced through by recourse to violence as the ultimate power resource. Neither is the system of representation one of party political competition, and when such competition does occur it is often regarded as dysfunctional.

The public sector and especially local authority organisational changes made during the 1980s and 1990s can be characterised as transfers of functions tied to different activities in the level-sector matrix. To the extent that transfers occur between the sectors changes also follow in management and coordination systems. Another type of change has meant that principles from another sector's management system have been borrowed. Gudmund Hernes (1980) notes such borrowings in all directions: internal price systems and corruption were market elements imported into bureaucracies; the market has imported competition controls and monitoring from the bureaucracy; consumer cooperation is an import from the association sector to the market sector etc.

Among the functions that have been transferred, special attention has been given to those of specification, financing, and provision (Savas 1982, Lundqvist 1991, Montin 1991). These can be illustrated by a local authority education department. In the traditional organisation it is financed by tax revenues. Specifications have been made by the politicians on the school board and the provision side has been taken care of by local authority schools. All three functions are public. Financing and specification are both central while provision is local. A transfer of political responsibility to a neighbourhood committee (*kommundelsnämnd*) or a local education board also transfers the specification function to the local level. In a system of pupil-linked school grants (service vouchers), where there are both local education and private schools, central financing is retained but part of the specification function is transferred to households at the private sector's local level. Provision can be found at all three levels, but especially at the local level.

We shall now attach names to these transfers: Transfers from the central to the local level are termed *decentralisation* and in the opposite direction *centralisation*. Functional transfers from the public to the private sector or adopting market mechanisms in the public sector are termed *privatisation*. Functional transfers from the private sector to the public or the borrowing of democratic/bureaucratic elements by the private sector are termed *socialisation*.

In the empirical analysis we will compare municipalities which have privatised or are in the process of privatising - in the form of functional privatisation or the introduction of market mechanisms into public sector activities - with those which, in the main, have refrained from doing so. The first category of municipalities we have termed *privatising municipalitites* the second *traditional municipalities*.

DEMOCRACY, EFFICIENCY AND JUSTICE

The different types of organisational change - decentralisation, centralisation, privatisation and socialisation - are justified in terms of three criteria: citizen influence, efficiency, and justice.

The main argument for decentralisation has been a democratic one: decentralisation increases citizen influence. The argument rests principally on the idea that it is reasonable for those who are most immediately affected to have a direct say in the activities which concern them. A geographic principle has often been applied to delimit the citizens concerned. Other democracy arguments have also left their mark on the doctrine of decentralisation: the small scale of the local level increases citizen involvement and the politicians' ability to manage. There is a relatively extensive discussion of the connection between scale and democracy which is far from conclusive (see, for example, Dahl and Tufte 1973). In favour of small political units is the greater 'weight' of each vote and the possibility for direct democracy. In favour of larger units, are economies of scale in service production and thereby more efficient democracy.

Alongside the argument for democracy, efficiency has also been invoked: local knowledge provides for better decisions. Likewise, the integration of different activities by virtue of the small scale enables the efficient utilisation of personnel and local resources (cf. Lundquist 1972).

Just as democracy and efficiency arguments have been presented in favour of decentralisation so too have they been used to argue for centralisation. The democracy argument for centralisation points to the difficulties encountered within modern complex societies in attempting to delimit those citizens most affected on a given issue and who ought thereby to have a greater say than others. In an integrated and complex society, all the parts are affected by all the others. All citizens ought therefore to have equal influence over all activities. The efficiency argument has, however, probably been seen as the most decisive. The central level is able to oversee the situation in a manner necessary for effective decision making. At the central level it is possible to avoid the sub-optimal politics which is characteristic of local decisions and which means that the collective good is reduced. Many actions demand expensive and specialised human and technical resources only available from the central level.

Alongside efficiency criteria the argument for justice figures as the most important reason for centralisation. In the matter of service provision, such as education, health care and social services, which is central for welfare states, it is considered to be vital that all citizens have access to the same quantity and quality of services. Within the exercise of authority such as taxation or the deprivation of freedom there ought not to exist any more extensive freedom of action for decision makers (cf. Regeringens skrivelse 1984/85:202, Davis 1968).

Privatisation is also justified in terms of the efficiency criteria and influence in the same order of priority. The efficiency argument builds, naturally enough, on market theory. Selfish competition ensures the most efficient conceivable distribution of production resources. But the market is also a system for citizen or, perhaps, rather, customer influence. The businessman who sells expensive and poor goods will see customers leave the shop to look for competitors. Those who do not adapt to customer demands go under in the market.

The argument for socialisation has once again stressed democracy. If democracy is a good system for citizen influence, then democracy in society increases as more activities and functions are placed under citizens' control.

Neighbourhood Committees and Market Orientation

Decentralisation and privatisation have been the key features of Swedish local government reforms and innovations during the 1980s and 1990s. The period of reform began with decentralisation. Behind this lay the consequences of the local government boundary reforms for democracy. The distance between elector and elected was believed to have increased. While a shrinking cadre of politicians (hereafter the '*political corps*') was experiencing difficulties in controlling a bureaucracy growing in size, expertise and strength. Conceived of as a problem for democracy, it seemed natural to achieve organisational changes through the implementation of decentralising measures in the interests of democracy.

Decision-making within representative bodies was to be moved from the municipal level to a newly created district territorial level. Neighbourhood committees (*kommundelsnämnder*, hereafter '*KDN*') were to be the instruments of democracy through electoral control of the politicians who, in turn, control the administrative services.

However, the territorial decentralisation of political decision making had several weaknesses. It suffered from decentralisation's classic conflict between democratic influence on the one hand and economic efficiency and justice on the other. Inevitably the solution to this contradiction is a compromise in which none of the interests are entirely satisfied. In the case of KDN the conflict was solved by means of central control in the form of indirect elections and by the allocation of funds from the municipal council.

Another conflict had its origins in nostalgic features of the KDN reform. The KDN reform was intended to increase the number of elected representatives with local ties, break down administration along sectoral lines, thereby strengthening the political level against the administration and create decision forums where local territorial loyalties would play a relatively greater role than party political and sectoral loyalties. One can discern the layman parish councillor as the ideal behind the KDN reforms.

Given that parish councils have long been unfeasible, the attempt to wind back the clock was doomed to failure. The compromise between central control and decentralisation was, perhaps, of a type that gave the centre too much domination over the periphery. During the late 1980s and early 1990s social developments showed new tendencies. Market liberals gained ground across a broad front, not only within Swedish municipalities, but even on a global scale (cf Olsen 1991). Citizenship preferences, it was argued, were better delivered by market mechanisms than through the hierarchical system of preference transformation comprised of the central and local state with its chain of democratic control. The increasing financial difficulties faced by local authorities gave rise to demands for more efficient allocation of resources and it was generally assumed that the market was better at this than democracy.

The question was no longer one of shifting the level at which decisions were made within the public sector. Instead, it became a matter of transferring decisions from the public sector to the market while market-like mechanisms were introduced into what remained of the public sector. Through a voucher system part of the specification functions could be privatised. The financing function could be privatised by levying charges. Production functions could be made more efficient through contracting out and the use of entrepreneurs. With the help of a purchaser-provider split model and various internal price systems, market mechanisms were to replace the ingrained democratic-bureaucratic management chain. Instead of a shift between levels a shift between sectors was now stressed.

Democracy, Representativeness and Legitimacy

Democracy can be conceived of as a system for the transformation of citizens' preferences into political decisions and actions which ultimately aim to affect the living conditions of the citizens themselves. Whether or not democracy can be judged to be successful ultimately depends upon the efficiency with which it is able to make these changes - its ability to deliver. This is an opinion found both in public debates and among political researchers.

Other commentators and researchers direct their attention to the processes which intervene between citizens' preferences and service delivery. Certain institutional arrangements have been considered to be of significance for democracy: universal and equal franchise, freedom of opinion, an informed and active citizenry, competition between political parties, a parliamentary relation between the legislature and the executive, and a successful control of the bureaucracy. The notion that politicians represent their constituents is also included.

The justification for taking note of what happens between citizens' preferences and the eventual delivery of services can, in principle, be justified in two ways: either one asserts that these intervening institutions and processes are prerequisites for an effective service delivery in accord with citizens' wishes, or one asserts that certain of these intermediate arrangements are either valuable in themselves or necessary for achieving other desirable ends.

As for the first justification, we can argue that only free elections between competing parties have been successful in delivering good living standards for the majority of citizens. Even enlightened autocracies have ultimately failed.

The second demands that different groups and categories of citizens have their own representatives among decision makers in the interests of justice. For example, arguments for an increased representation of women have deployed both justifications (Hernes 1987). The representation of women has been presented as a resource for decision making and a necessity given that women have different interests than men. If women do not themselves take part in the decision process then these interests will never be expressed. Both of these arguments mean that the political system's ability to deliver increases if women are represented. The second type of justification argues that it is unjust if women are not represented.

Politicians as 'listening posts' whose job it is to pass on the electoral group's wishes and thereby achieve better results. The problems which arise when listening posts decline are not only confined to Sweden; they have also been noted in Norway (Larsen and Offerdal 1992) and in Great Britain. Two prominent British local government researchers write as follows:

> There are dangers in any reduction in the number of councillors if importance is attached to the representative role of councillors (Clarke and Stewart 1989).

The connection between the number of politicians and the representation of minority groups has also been empirically established in different political systems. Alozie and Manganaro (1993) show, for example, that the likelihood of blacks and Hispanics being represented in American cities increases along with an increase in the number of councillors.

If the system is unable to deliver or does not satisfy the criteria which are judged necessary for efficient delivery, or if the system does not satisfy the criteria deemed important for other central values then the risk exists that citizens will no longer regard the system as justified. The political system's legitimacy is called into question. Legitimacy can be tied to the ability to deliver, to parts of the process which are understood to be vital for its ability to do so, or are important for the achievement of other valuable goals like justice, equality and the possibility to exert influence.

Montin (1993), drawing upon Weber (1947) and Offe (1985), discusses sources of legitimization for organisational changes in Swedish municipalities. His point of departure is the distinction between *formal-legal rationality* and *functional rationality*. This distinction closely resembles that between the concepts ability to deliver and process the criteria discussed here. Montin notes that:

There is clearly a conflict between the market orientation and the formal-legal model.

The entire idea that different groups and categories having their own representatives is of value to democracy has been challenged by researchers and commentators who argue that it is not the group to which representatives belong but their opinions which is decisive (cf. Holmberg 1974). One can also go a step further and assert that it is not the representative's opinions which are important but only the results of the political process. The relationship between the wishes of citizens and policy output has been termed *responsiveness* (Anckar 1980). If responsiveness takes the place of representativeness, then one has aligned oneself with a view of democracy in which process criteria are no longer significant compared to the system's ability to deliver.

Representativeness is clearly understood as just such a process criterion. The electoral system is based on the principle of representing the territorial constituencies and the different opinions of the populace. The political parties' nominations process devotes considerable attention to the representation of social categories and interest groups. Many groups are themselves active in promoting their own representation. Regardless of whether or not representativeness is a necessary criterion for the system's ability to deliver, its legitimacy will become shaky if representativeness is weakened.

Democracy and citizen influence are cited as justifications for both political decentralisation to KDN and for the move to privatisation and market orientation in local authorities. Like other systems, democracy is dependent upon legitimacy for its survival. Legitimacy can be affected by the system's degree of representativeness. This is the normative justification for our empirical investigation. The results are presented below. Naturally it must be stressed that the conclusions drawn apply *ceteris paribus*. If one or the other organisational change - decentralisation or privatisation - can be shown to have negative consequences for representativeness (and thus in all likelihood the system's legitimacy) then the conclusion can be drawn that the actual change has failed only if other components of democracy and effectiveness have not been developed so as to compensate for the loss of representativeness.

EMPIRICAL MATERIAL

Swedish political scientists have twice (in the mid 60s and in the shift between the 70s and the 80s) carried out large local government research programmes intended to evaluate the consequences of the local government boundary reform. Strömberg and Westerståhl (1984) summarise these previous research programmes. A third programme, focusing upon the effects of subsequent changes in local government organisation and milieu on democracy, was initiated during 1991.

In all three research programmes parts of the investigation have been carried out in a selection of local authorities. In the latest of the completed programmes - the Local Government Democracy research group - a randomly selected sample of fifty local authorities was studied. Those parts of the studies relevant for our present concerns were carried out in a randomly chosen sample comprised of half of the original fifty local authorities.

When the sample of local authorities was constructed for the current research programme - Democracy in Transition - it was natural to begin with the Local Government Democracy research group's 1979 sample. The sample was supplemented with a sample of strategically chosen county councils. Some local authorities were replaced and others added, partly to insure the presence of a fairly large number of local authorities with KDN, and partly to capture a reasonable number of primary local authorities (the first tier of the two-tier local government system) within the selected county council areas. The earlier research programmes excluded large cities from the research population. In order to tackle the special problems of metropolitan authorities, it was decided to include the City of Göteborg in the sample (Johansson, Lorentzon and Strömberg 1993).

The final sample consists of 28 local authorities and four county councils. Twenty of the 28 local authorities were also present in the 1979 sample of 25 local authorities. Given that measures relevant for the current inquiries were made in 1979, it was natural to carry out this investigation within the framework for the research programme which was already under way using the 28+4 local authority sample. The investigation can be regarded as a panel investigation for 20 of the local authorities. This increases the possibility of saying something about changes over time. For the remaining eight authorities (and the four county councils), we can only give a picture of the situation in the Autumn of 1992. Hints are present of how the local authorities, at the same point in time, differ from each other with respect to the organisational reforms.

Lists of elected representatives from the 32 local authorities (including the county councils) have been drawn up. The information from these has then been compiled into personal registers. The information which is regularly found is assigned structure (the tasks a person has) party membership, sex and, in exceptional cases, profession and/or age.

The material has subsequently been sorted by local authority and political party and the lists handed over to local party organisations for completion. In addition to the information on age and profession we have also requested information about elected representatives born abroad, as well as in which parts of the municipality they live (central residential districts or countryside). The latter information has not been collected from the county councils. The response frequency has been relatively satisfactory. Of the 231 local party organisations 194 (84 per cent) answered our questions. In the analysis of representativeness ,use has been made of public statistics.

The information gathered in this way concerns the size of the elected representatives group and the population composition which is directly relevant for the descriptive questions: How has the size and representativeness of the political corps changed between the two measurement occasions? A more analytical aspect of the question is, however, the significance of the organisational changes. We must therefore acquire some idea of the actual organisational models present within the local authorities.

The easiest local authority model to distinguish is that of a blanket organisation of *KDN*, i e a neighbourhood committee organisation covering the whole territory of the authority. There are four of these in our material. It is more difficult to determine where the more market oriented model has made headway. Three studies have provided us with pointers: The National Association of Local Authorities post electoral study of 1991 indicates where the purchaser-provider split model has been or is planned to be introduced; the Local Democracy Committee's local authority 1992 inquiry provides information on the occurrence of profit centers, private entrepreneurs, co-operative or associational entrepreneurs, newly established municipally-owned companies and expert committees. A questionnaire sent to local authorities in 1992 as part of an evaluation for a new state grant system for refugee reception programme, provides information on delegation, privatisation and management by objectives (Bäck and Soininen 1993). After having weighed the results from these studies, we decided to regard nine of the elected local authorities as advanced cases of privatising organisations. The results show the local authorities have been categorised into one of the two models described above. The county councils are regarded as a special category as is Göteborg.

How Many Politicians?

A direct comparison between the different report groups with respect to the number of politicians can easily mislead. It has been known for some time that the total number of elected representatives varies along with population size. Our five result groups are not randomly comprised of local authorities of the same size. Another point which should be noted is that the connection between population size and number of politicians is not proportional. certain 'minimum organisation' exists. This means that 'politician density', i.e. the number elected per inhabitant, falls with increased local authority size. Politician density as a relative measure can lead us to draw incorrect conclusions.

Although the situation is not entirely ideal from a statistical point of view (relatively skewed distributions), we have chosen to use the residuals, i.e. deviations from the regression line *Politicians total = A + B x Population,* as a comparative measure.

Table 1. Average number of politicians in different report groups (whole sample of authorities)

Report group	Number of politicians[1]	
	Mean	Mean deviation from predicted number
Traditional organisation	162	-1
KDN organisation	323	+123
Privatising authorities	148	-21
Göteborg	705	+297
County councils	231	-123

[1] Includes ordinary and deputy councillors as well as ordinary and deputy members of committees (selection of committee members in Swedish local authorities is not restricted to the members of the council)

In Table 1, the results of the average comparison totals for the report groups are given. The traditionally organised local authorities hardly deviate at all from the expected values. However, strong positive deviations can be noted both for Göteborg and the three local authorities with blanket KDN organisations. The county councils show a strong negative deviation and the group of privatising local authorities a less marked but still clear fall. A tentative conclusion from the cross-section analysis is that the organisational form of KDN increases the number of elected representatives, while the privatising organisational form tends to lead to a decrease in the number of politicians. We shall now consider changes over time to see if this tentative hypothesis withstands more rigorous testing.

Data from the National Association of Local Authorities with regard to changes in the number of *political posts* are in line with these results. These results can be summarised as follows. The total number of political posts decreases in all local authorities, but the reduction is most notable in those with a privatising organisation. The sharp decline in the political corps in these local authorities can not entirely be explained in terms of the rationalisation of the appointments structure. It is also clear that increased recruitment difficulties have arisen in those authorities which have introduced these more recent organisational arrangements. The parties are facing the increasing problem of finding willing and able candidates for office.

Table 2. Average number of politicians in the 'panel authorities' 1979 and 1992

Report group	1979	1992	Percent change
Traditional authorities	185	163	-12
Privatising authorities	188	129	-31
Effect of privatisation			-19

We have no empirical material with which to explain the increasing recruitment difficulties. If one wishes to look for an answer, it is reasonable to see the problem in terms of 'will' and 'ability'. Is it that the new organisational form has severed the direct channels of control between politicians and the bureaucracy thus making the job of politician unattractive for presumptive politicians and hence difficult to fill? We have known for some time that it is precisely the desire to influence that is the true driving force behind the recruitment of elected representatives and the factor that keeps them within the appointment system (Wallin, Bäck and Tabor 1981).

Or is it the case that the new types of appointments have become too difficult? Does the purchaser role place such large demands on the elected representatives that they are unable to manage the job?

If the number of office holders decreases more than the number of appointments (in council and committees) then the concentration of duties increases with mathematical certainty. The concentration of duties - number of duties per person-was calculated in the 1979 study to be 1.4. In the

National Association's 1977 post electoral study the figure was also 1.4. No figure is available from the National Association after this date. Our results suggest an increase. The mean value across the 32 local authorities is 1.53. There are, though, significant variations between the different report groups, the Göteborg figure is 1.28 and the county councils' 1.75. The three organisational types can be ranked : (1) privatising 1.56 (2) traditional 1.51 and (3) KDN authorities 1.34. The tentative conclusion to be drawn is that the introduction of the KDN organisation is associated with a reduction in the number of politicians with multiple duties. The reverse applies with the introduction of market oriented organisation.

Precisely, as in our earlier analysis of changes in the size of the political corps, we will now see if these conclusions are compatible with the results over time. Because the 1979 study did not include those who were deputy council members we have filtered out those with deputy duties from our 1992 data. Neither, of course, have we counted posts as a substitute or deputy on councils as a numerator in the quota we term concentration of duties.

It is evident from Table 3 that the number of politicians with multiple duties has increased in both study groups. The increase is especially noticeable in those authorities that have introduced a more market-oriented organisation.

Table 3. Average concentration of duties. Panel Authorities.

Report group	Duties/person		Change
	1979	1992	
Traditional authorities	1.39	1.43	+0.04
Privatising authorities	1.38	1.51	+0.13
Effect of privatisation			+0.09

Who Are the Politicians?

The representation of different groups varies quite substantially depending upon the kind of political post we consider. A number of observations can be made when we examine the pattern of representation across different types of organisations and posts. A striking observation is that several of the under represented groups are better represented as substitutes or deputies in specific organisations than as ordinary members. This relationship applies in seven out of eight cases for women, in eight out of eight for workers and 'the young', and in five cases out of eight for immigrants. Clearly a question of status is involved here. Members of more underprivileged groups can acquire posts as low status deputies more easily than as ordinary members.

However, for two of the under represented groups - 'the old' and central district residents- this relationship does not hold. In seven out of eight cases these two groups are better represented among the ordinary members than the deputies. Under representation among the older and central district residents is clearly not tied to status in the same way as for women, workers, immigrants and youth.

It can also be noted that it is precisely these groups - women, workers, immigrants, and the young - who are consistently and weakly represented in the most prestigious of the committees - the executive board.

A measure of representation which will allow comparison between different under and over represented groups and between local authorities must take account of the groups' varying population size. We do this by calculating a representation index which expresses over and under representation as a proportion of the group's share of the population. More formally:

$$RI_x=(p_{px}-p_{bx})/p_{bx}$$

where RI_x is the index of representation for the group x,Pp_x and Pb_x are the group's share of the political corps and population, respectively. This index will be 0 when a group has a share of the political corps that corresponds to its proportion of the population. Positive values denote over representation and negative values under representation. A group in the population totally without political representation will have the index value -1.

In Table 4 these measures are presented for citizen groups and for the different local authority study groups. In the comparison of the different organisational models, the KDN organisation appears to deviate when it comes to the pattern of representation. Workers are strongly under represented in KDN authorities while the old and central district residents enjoy relatively good representation. These results are difficult to interpret. The variations probably originate in an initial and unexamined situation and are so large as to resist interpretation in terms of the effects of organisational changes.

Table 4. Representativeness index for different groups. Means per report group

Report group	Women	Workers	Young	Old	Immigrants	Central area
Traditional author.	-0.26	-0.41	-0.79	-0.49	-0.66	-0.15
KDN organisation	-0.20	-0.71	-0.71	-0.21	-0.67	-0.04
Privatising author.	-0.25	-0.49	-0.80	-0.42	-0.50	-0.09
Göteborg	-0.22	-	-0.83	-0.35	-	-
County councils	-0.09	-0.67	-0.83	-0.44	-0.65	-
All municipalities	-0.23	-0.50	-0.80	-0.43	-0.65	-0.12

We shall therefore turn to studying the changes over time. As in earlier analyses of this type, we lose the opportunity to say anything about changes linked to the introduction of the KDN organisation. We must content ourselves with conclusions pertaining to the introduction of a more market-oriented organisation.

A contribution to the study of how representative KDNs are can be found in a small study from three KDNs in Stockholm (Premfors and Sanne 1990). The authors draw the following conclusion:

> If we specify a norm that politicians should be entirely representative and perfectly reflect the population... then the new KDNs clearly fail the test. On the other hand, if we establish a norm which proceeds from a comparison with other political decision-making assemblies then they must equally clearly pass.

There is no information from 1979 of the population distribution across socio-economic groups. In order to have some kind of reference point, we have related the proportion of working class politicians to the proportion of the working population employed in manufacturing and the construction industry.

The representation of women improves in both study groups, but most in the traditional organisations. In an analysis analogous to that made of changes in the number of politicians and changes in concentration of duties we can calculate the effect for women's representation by introducing a market oriented organisation as being 0.15 - 0.19 = -0.04.

The representation of the working class has also been improved according to our measures. This improvement is above all dependent upon deindustrialisation. The average proportion of politicians who we have classified as workers remains largely unchanged in the 20 panel authorities, while the proportion of the population who are employed in industry declined dramatically from 43 per cent in 1979 to 33 per cent thirteen years later. In all likelihood the proportion of the population who are working class has not fallen as sharply as industry employment. The improvement in working class representation is therefore presumably less marked than our figures are able to show. The fact remains, however, that, to the extent that working class representation has improved, it has done so in a different way than for women. The improved representation of women has come about through an increase in the proportion of female politicians, while working class representation has improved because the working class, while shrinking in size, has maintained its total number of elected representatives.

Table 5. 1979 and 1992 representation index in the panel authorities. Means per report group.

	Traditional organisation			Privatising authorities		
	1979	1992	Change	1979	1992	Change
Women	-0.45	-0.27	+0.19	-0.43	-0.28	+0.15
Workers[1]	-0.36	+0.05	+0.41	-0.38	-0.13	+0.25
Young	-0.53	-0.79	-0.26	-0.53	-0.87	-0.34
Old	-0.56	-0.50	+0.06	-0.62	-0.47	+0.15
Immigrants	-0.76	-0.68	+0.08	-0.48	-0.59	-0.11
Central	+0.14	-0.11	-0.25	-0.06	-0.02	+0.04

[1] The reference group in the general population is the proportion of the residential workforce who are employed in industry and construction.

As in the case of women representatives, the improvement for the working class is most noticeable in local authorities with a traditional organisation. The effect of introducing a market oriented organisation can be calculated as 0.25 - 0.41 = -0.16 i e a deterioration in the representation of the working class by 16 percentage points.

Representation of the young and immigrants follows a similar pattern: The representation of the young is worsened and to a greater extent in the privatising authorities. The representation of immigrants improves in traditional authorities but less in the privatising. The effect can be measured as -0.08 for the young and -0.19 for immigrants.

When we consider the remaining categories- the old and-residents in central districts - the effects are in the opposite direction. The effect of privatisation on the representation of persons over 60 years old is +0.09 and for central district residents +0.29. We may recall from the introductory account that these groups both deviated from women, workers, immigrants and the young in the way that they tend to be relatively well represented in more prestigious organisations and duties.

Organisational Changes and Democracy

The organisational changes within local authorities during the 1980s and 1990s can be described as decentralisation succeeded by privatisation. Both aims have been justified in terms of citizen influence and efficiency. Decentralisation has particularly stressed the goal of increased influence, and privatisation that of efficiency. The influence of citizens is channelled differently within decentralised as compared with privatised organisations. Decentralisation envisages citizens as exerting their influence through the democractic-bureaucratic chain of command. The system of representation, centred upon the electoral system, is the major institution. After privatisation, market mechanisms act so as to mediate the influence of citizens. According to theory, voters make decisions within the democratic management chain through comparing their own preferences with the alternative futures offered in the party programmes. Demand in the market is the product of the preferences and resources of citizens.

In a system of representative democracy politicians play a role in the survival of the system either directly or through legitimacy. In this paper we have tried to establish the effect decentralisation and privatisation have on political representation in local Swedish authorities.

The results are fairly clear and can be summarised:

1. A decrease in the number of politicians is closely connected with the introduction of market arrangements. In the long run, the introduction of a blanket KDN organisation leads to an increase in the number of politicians.

2. The number of politicians with multiple duties has also increased, and here too the introduction of market-oriented organisations is of great significance. The introduction of the KDN organisation has probably had a restraining effect on the increase in multiple duties.

3. The introduction of privatisation has worked against the representation of deprived groups. It has not been possible to decide if the introduction of a KDN organisation has had positive effects in this respect.

We have established that this development can be a threat to the legitimacy of the system. A special problem is that the Swedish political system, which rests upon a structure comprised of the political parties and interest organisations and processes such as the electoral system and the system of representation, is already suffering from a weakened legitimacy. Several studies clearly point in the same direction. Namely, that the Swedish people's loyalty and faith in the system and its institutions is failing (Bergström 1991; Bäck 1992; Elliot 1993; Petersson et al. 1989). It may therefore appear ill-considered to expose the system to the further strains implied by poorer representation.

References

Alozie N.O. and Manganaro L.L. (1993) 'Black and Hispanic Council Representation: Does Council Size Matter?' *Urban Affairs Quarterly* 2:276-298.

Anckar D. (1980) 'Responsivitet - Om begreppet, företeelsen och forskningsområdet' *Politiikka*l.

Axelrod R. (1987) Från konflikt till samverkan. *Varför egoister samarbetar* Stockholm: SNS förlag.

Bergström H. (1991) 'Sweden's Politics and Party System at the Crossroad'. In Lane J E (ed.) *Understanding the Swedish Model* London: Frank Cass.

Bäck H. (1990) *Att vilja, att kunna, att böra. En studie över statens styrning av den kommunala ekonomin.* Stockholm: University of Stockholm, Department of Business Administration, Institute of Local Government Economics.

Bäck H. (1992) *Politiskt ansvar och kommunförvaltningens styrbarhet. Om partipolitisering, målstyrning och marknadsstyrning* Stockholm: University of Stockholm, Department of Business Administration, Institute of Local Government Economics.

Bäck H. and Soininen M. (1993) *Flyktingmottagning i förändring* Stockholm: Swedish Association of Local Authorities.

Clarke M. and Stewart J. (1989) 'The Future of Local Government: Issues for Discussion' Birmingham: University of Birmingham, Institute for Local Government Studies (mimeo).

Dahl R.A. and Tufte E.R. (1973) *Size and Democracy* Stanford: Stanford University Press.

Davis J.W. (1968) *Little Groups of Neighbors: The Selective Service System* Chicago.

Elliot M. (1993) 'Medborgerlig tillit och misstro. Om samhällsförtroende i kristider.' In Holmberg S and Weibull L (eds.) *Perspektiv på krisen* Göteborg: University of Göteborg.

Hernes G. (1980) 'Forhandlingsökonomi og blandningsadministrasjon'. In Hernes G (1980) *Makt, forhandlingsökonomi og blandningsadministrasjon* Oslo: Universitetsforlaget.

Hernes H. (1987) *Welfare State and Woman Power - Essays in State Feminism* Oslo: Norwegian University Press.

Holmberg S. (1974) *'Riksdagen representerar svenska folket.' Empiriska studier i representativ demokrati* Lund: Studentlitteratur.

Johansson F., Lorentzon L.O. and Strömberg L. (1993) *Kommunmedborgarundersökningen 1991. Undersökningsdesign, urval, datainsamling, marginalfördelningar och bortfall* Göteborg: University of Göteborg, Department of Political Science.

Larsen H.O. and Offerdal A. (1992) *Demokrati uten deltakere? Arbeidsvilkår og lederroller i kommunepolitikken* Oslo: Kommuneforlaget.

Lundquist L. (1972) *Means and Goals of Political Decentralization* Malmö: Studentlitteratur.

Lundquist L. (1991) 'Privatisering - varför och varför inte?' in Rothstein B (ed.) *Politik som organisation* Stockholm: SNS Förlag.

Montin S. (1991) 'Privatiseringsprocesser i kommunerna' Örebro: University of Örebro (mimeo).

Montin S. (1993) *Swedish Local Government in Transition: A Matter of Rationality and Legitimacy* Örebro: University of Örebro.

Offe C. (1985) *Disorganized Capitalism* Oxford: Polity Press.

Olsen J.P. (1991) 'Rethinking and Reforming the Public Sector' Bergen: LOS-senteret.

Petersson et al. (1989) *Medborgarnas makt* Stockholm: Carlsons.

Premfors R. and Sanne M. (1990) 'De nya stadsdelspolitikerna.Social representativitet och lokal förankring'. Stockholm: University of Stockholm, Department of Political Science (mimeo).

Regeringens skrivelse 1984/85:202 om den offentliga sektorns förnyelse.

Savas E.S. (1982) *Privatizing the Public Sector* Chatham: Chatham House.

Sharpe L.J. (1989) 'Fragmentation and Territoriality in the European State System' *International Political Science Review* 3:223-238.

Smith A. (1776) *The Wealth of Nations* (*Der Wohlstand der Nationen)* München: Verlag C H Beck (1974)).

Strömberg L. and Westerståhl J. (1984) *The New Swedish Communes: A Summary of Local Government Research* Stockholm: Liber.

Wallin G., Bäck H. and Tabor M. (1981) *Kommunalpolitikerna. Rekrytering - Arbetsförhållanden - Funktioner* Ds Kn 1981:17-18. Stockholm: Ministry for Local Government.

Weber M. (1947) *The Theory of Social and Economic Organization* New York: Free Press Paperbacks (1964).

Zagare F. (1984) *Game Theory. Concepts and Applications* London: Sage.

Chapter 6

Community Councils - Improved Local Democracy?

Allan Dreyer Hansen, Torill Nyseth
and Nils Aarsæther

1. Introduction

With an obvious decline in citizens control and influence on public decision-making through the "parliamentary channel", some efforts have been made to restore this influence and control by creating new institutions and channels of influence. The result of such reform efforts are typically found at a local level, often in the form of "user-democracy" for people being served by different public institutions. However there has also been some attempts at delegation or decentralization of political power to neighbourhood and sub-municipal levels, so-called neighbourhood councils, or community councils (CC).

Hope has been expressed that such arrangements would lead to a more democratic way of regulating public affairs (J. Andersen et al., 1993: 236). An evaluation of type of CC's found a possibility for the emergence of a new and more responsive type of politician among the CC leaders (M. Anderson et al., 1989: 21 and 279).

The purpose of this chapter is two-fold. First, we shall discuss different criteria for evaluating the political democratic gains from territorially (as opposed to functionally) based citizen influence. Second, our (provisional) results from studies of CCs in Norwegian and Danish municipalities will be discussed in terms of these criteria.

The first part of the chapter will focus on the following questions:

- Are community councils to be understood in terms of direct or representative democracy?
- If we suppose that the councils actually have some kind of political power, we ask whether this power is increasing the democratic aspect of the decision making process on the community as well as on the municipal level. CC's can be said to increase democratization in four ways: First, they may stimulate the local political debate or in other words function as an institutionalized local public sphere. Second, they may facilitate citizens participation in community-oriented issues. Third, they may influence the agenda-setting both at the community and the municipal level in a way which might better correspond to local opinion-formation than was formerly the case. Fourth, they may function as a channel of recruitment to the higher levels in the political system.

The second part will be a presentation of the four different types of neighbourhood councils and their relation to the municipal level, in the municipalities of Tromsø and Stavanger in Norway, Ejby and Herlev in Denmark. Here we shall discuss the different forms of council arrangements in relation to the above mentioned criteria of democratization. In the third part we will analyze our results in the four cases will be analyzed in a comparative perspective.

1.2. What Do We Mean by "Improved Local Democracy"?

When dealing with the question of an extension for the democratic element in politics and in society at large, the need for certain standards or criteria for what is meant by "democracy" is obvious. Discourses on democracy often begin with a focus on the distinction between indirect and direct forms of democracy (cf. A. Ross, 1967; H. Koch, 1981; D. Held, 1990). This is, however, a matter of degrees, not of absolute and definite types. Almost nobody will support the idea of direct democracy as the only legitimate way of handling matters of common concern. The idea of *demos*, coming together and discussing every issue relevant to the community is no longer conceived of as viable. Rather, advocates of radical democracy argue in terms of representatives with mandates given by the voters, short election periods, and a frequent use of referenda. On the other hand defenders of indirect democracy never speak about life-long offices. They advocate fixed election periods, but will defend the right of the representative to vote and act independent of the electorate.

The direct - indirect dimension is, however, only one important aspect of democracy. It should be combined with a dimension of individualism vs. collectivism. In terms of political theory, this distinction means to what extent one conceives democracy as a means of aggregating and balancing the individual, pre-political preferences of the citizens. This is in contrast to the conception of democracy as a means of arriving at a common will or a shared conception of the common good by the active involvement of citizens in the process.

Normally one supposes that "collectivists" will be the advocates of direct democracy, while "individualists" will prefer a system of representation. But this is not necessarily the case. A. Ross is by principle an advocate of indirect democracy. He argues that the representatives of the people (better than the people themselves) will arrive at the best solutions in accordance with the will and the interest of the people. It could also be argued that a direct democracy is perfectly suited to the promotion of individual preferences.

It is our contention that a democratic system will stagnate if the only action on the part of the people is to choose between representatives every four years. On the other hand, a system of direct democracy can only be expected to function in small scale communities. The frequent use of referenda or a continuous participation by the means of electronic media is of course possible, but will probably tend to render the citizens atomized; responding to, but not in the position of initiating issues. When we consider Community Councils, we are primarily concerned with an increase incitizens participation in matters of common interest.

One last point to be discussed is the question of levels in a democratic system, the relation between local and central levels of authority. Tocqueville contended that local democracy is *natural*. Not challenging his position, we nevertheless have to add some words along the way. To start with, the question of local democracy is one of balancing the value of *freedom* at the local level by the value of *equality* as a national or central level concern. To day, at least in the Nordic welfare democracies, this possible conflict between the two values is no longer the most important characteristic of the central - local relationship. It is rather a question of common government. However the local level is not completely reducible to (functionally based) implementation of ends formulated at the central level. It is rather a question of frames or minimum-standards inside of which, the local level enjoys a certain autonomy.

There are at least three principal reasons why in certain fields such a local autonomy should be given more weight than national or central level equality. First, the limits of a demos should correspond to the area in which externalities could be traced. This means that only those who are affected by a decision - directly or indirectly - should be the ones that influence it. Obviously, there are many decisions that have no or negligible effects outside the local community in question. Second, there are limits to the monitoring capacities of central institutions. Already by 1946, Alf Ross suggested decentralization in order to avoid "bureaucratic absolutism" (A. Ross, 1967: 246). The third argument is the question of learning democracy by practising it. This last point has always been of central interest to the theorists of participatory government (Koch, Pateman etc.), but Ross also stressed this side of the case for decentralization (Ross, 1967: 86).

As we have pointed out earlier, comprehensive political participation can primarily be achieved at the local level.

The next question is why a combination of municipalities and sub-municipal community councils is preferable to a system of (small scale) municipalities. We have no intentions of entering a debate on the question of the ideal size of a municipality. Such a debate will most probably have no practical implications, as the municipal structure in each nation-state will have to be understood and handled in a complex context of geographical, historical and political dimensions. We will take the municipal structure in Denmark and Norway more or less for granted. Furthermore, we will run the risk of maintaining that the municipal democracy to a certain degree has stagnated, perhaps more so in Denmark than in Norway (Togeby, 1992). Relatively few people actively participate, several political parties report recruitment problems, and only a minority of the people knows what goes on in the municipal council.

The question we will try to answer is whether a system of CCs, given the present municipal structure, can vitalise democracy at the local level. We will discuss this question with reference to the above mentioned criteria, which will be developed in the following sections.

1.3. Central Principles

With reference to our discussion in the first part of this paper about the idea of democracy, let us now turn to the CC and discuss in what way the CC can be said to increase democracy. We have already sorted out at least four criteria or principles of democratization.

The first one is the CC`s ability to *institutionalize a local public sphere.* First of all this would be expressed by an intense and open local political debate, due to the citizens access to a new channel of political communication and influence. Such a strengthening of the political debate might be accomplished by the creation of formal structures; that is, specific rules and procedures of communication, and through the business of the CC, that is, what kind of issues are allowed to enter the local agenda. If the CC just adopt the rules and procedures known from other units, for example the municipal council, it is also possible that the debate on the community level will suffer from lack of inspiration and creativity to the same degree as the superior unit. It is obvious that the CC agenda will determine who and how many citizens will participate. We suppose that the local agenda first of all will be dominated by community issues. But that does not exclude other issues of broader interest, not even questions of an international range. The CC policy, however, will normally be dominated by local interests. "Community issues" might even be given another meaning in the CC context than through, for example, the local authorities.. Local knowledge, the articulation of local interests and preferences etc. will to a higher degree be considered suitable in the work of the CC.

The second criterion is the CC`s potential to *mobilize the citizens in the communities.* This might be the case if the citizens interest and engagement in local issues are more visible than before, after the introduction of the CC in a community. The degree of mobilization can be measured by the ratio of the inhabitants participating in the elections to the CC council, but this will only be relevant to CC councils being elected directly by the citizens. When elections take place in an open meeting, not just using a secret ballot, one will get fewer participants. At times all councils, irrespective of election or appointment methods, will arrange meetings open for all inhabitants of the community. Mobilization can then be measured by counting the attendance on such occasions. And the same goes for other occasions or issues that the CC initiates, demanding popular participation

- "dugnad",[1] work of different types etc. The ability to mobilize is also expressed in stability and in the breadth in the recruitment to the CC.

The third criterion questions the CC`s success in *introducing a new channel of "bottom-up"* influence in the formal political system that in one way or another influences the agenda of the municipality. A premise, however, is that the CC should have some sort of discretion. This might be the transferring of some specified competence, but also budgetary resources for free disposals, and even forms of access to the municipal decision making process. If the CC's have legal authority their decisions should not be questioned by the local authority. Or, if the CC's have not achieved formal authority of some kind their opinion should at least be registered and reported to the municipal councillors, thus making CC influence possible.

The fourth principle is to what degree the CC *expands the sphere of political participation* in such a way that the participation in general increases, not just transfers from one level to another. If the election to the CC causes a competition between the CC and the municipal council, in such a way that the interest in running for the local council is diminished, we have an undesirable zero-sum situation. Connection between the levels, however, is needed for example, in such a way that some sort of political career pattern can be achieved. The CC might for example function as a recruitment channel for a political career at higher levels. In this way, the CC might be seen as a place for learning democracy. This was very much in the thoughts of J. S. Mill.

The CC might also represent a correction towards the central political culture, especially in regard to problem solving by using other and more experimental and problem-oriented methods and strategies. In this way it contributes to a vitalizing of the political culture in general.

To summarize, the CC system should lead to an increase in democratization, both in quantity and quality, especially on the community level, but also on the municipal level of the political system. Some problems, however may arise.

1.4. Possible Dysfunctions of CC`s

The CC may be susceptible to misuse and even have an adverse effect on democracy at large.

---

[1] "Dugnad" is a Norwegian expression hard to translate to English, but means the institutionalization of collective, unpaid work on a project in a community, in principle based on voluntary participation, but with informal sanctions against non-participants.

CCs may fail in mobilizing the local public, or may mobilize only certain segments of the local population. Often one would think of such problems in terms of electoral arrangements, and of course electoral arrangements may enhance or prevent popular support and mobilization. But the problem of establishing truly democratic CCs may arise independent of electoral arrangements. We can identify a bottom-up impediment to democracy, as well as a top-down one.

If the people in the area are empowered to elect their CC, by ballot or in an open meeting, we will expect that the procurement of candidates will not normally involve political parties. An electoral committee will often present just one list of candidates prior to the election, and alternate candidates will not normally be running for office. Individuals with a local standing, not far from the informal centres of authority at the local level, are likely to appear on the list of candidates. They will seldom be affiliated with a political party. In the extreme one can run the risk of having elected a council of local honoratories, obviously not a democratic gain, but in perfect accordance with the rules. So far, we have had no opportunity to study to what extent individuals or groups feel excluded from election preparation.

But if the people living in an area must accept a CC appointed by the municipal council, with partisan representation corresponding to that of the municipal election results, the relation between council members and the local public is critical. The citizens of the community are simply not allowed to get rid of unpopular councillors in the CC by electing others. An appointed CC will lack a local political legitimacy, and the political role of the public in the area will be unsatisfactory. If the CC officers are not responsive to local opinion, people with grievances must approach the elected representatives in the *municipal* council.This body may or may not instruct the CC to pay more attention to local points of view.

In principle, the actions of the CC should reflect the views developed among the citizens of the local community. This presupposes the existence of arenas for the formation of opinions on issues of common interest. Neither the election occasion (if a CC is directly elected) nor the open access to CC meetings will provide sufficient occasions for this opinion formation. Without some public arenas for formulating and debating common issues, and a connection between these arenas and the CC, the deliberations of the CC will not necessarily reflect the concerns of the demos involved. These arenas for the formation of public opinion and the contact between the CC and the people have been organized as popular meetings, or as the citizens' and the councillors access to written and transmitted local media. More informally, debates can take place in local meeting places or occasions attended by a large or at least a substantial part of the population, occasions or places that are in principle open to all.

If we suppose that the council *is* representative of the people living in an area, and that the deliberations in the CC reflects the local opinion, the functioning of a CC system may nevertheless be counter-productive for democracy. Two dangers are obvious. First the NIMBY (Not In My Back Yard) syndrome, exemplified by the efficient and representative CC that succeeds in gaining special favours from the municipality by out-manoeuvring people or CCs in other geographical areas within the municipality. CC activities promoting myopic action on behalf of a community's residents will be an obvious threat to a well-functioning municipal government system.

Secondly, a system of representative and enlightened CCs, not NIMBY-infected, may indirectly pose a threat to the local democracy and even to democracy at large by being too attractive. As an effect, political parties lose support and recruitment, and this again may affect the democratic element in the governing of municipalities. The "official" municipal council may be drained of motivated persons as representatives, to the extent that people of this kind find CC politics more attractive than holding "real" local government positions. A possible effect of such a process may be that local level administrators take over more of the municipal decision-making.

This will be the case when the well-functioning CC suffers from lack of competence, and in a situation in which it is not only neglected, but even deliberately by-passed by the municipality in matters of definite relevance to the *demos* in question. A study of a CC in the municipality of Drammen (near Oslo) reported that the CC representatives left their offices in protest; the municipality had avoided contact with the CC and instead established direct contact with a local business organization in developing a new road project (Carlsson, Y., 1992). The continuous functioning of a CC devoid of discretion and competence is unlikely, and as they disappear, CCs without power will cease.

Efficient and empowered CCs may represent a problem: A gathering of efficient community organizers and entrepreneurs around the CCs, combined with a weak political leadership at the municipal level may be a nightmare for personnel working at the higher level. Not all communities or residential areas of the municipality are likely to be endowed with successful CC people. There is a small chance that the aggregate of CC initiatives will produce more equality in living conditions and welfare levels across the municipal territory.

2. Four Cases

In this part we shall present our provisional results from our case-studies in four municipalities in Norway and Denmark. The presentations will be somewhat more complete in the Tromsø case than the others because there the field work has just begun
. The presentations will end with some concluding remarks on the main questions in this chapter.

2.1. Stavanger (Norway)

The urban municipality of Stavanger is situated in the South-West part of Norway. The city is the regional centre of the region Rogaland. The municipality covers quite a large area of communities or districts surrounding the city. For several years, there neighbourhood councils have existed in 11 of these areas. These, however, have been endowed with very little formal authority. In 1987, Stavanger joined "Frikommuneforsøkene", a national program for modernizing the local government structure in Norway (Rose, L., 1991). The contribution from Stavanger to this programme was an experiment of decentralizing power and formal authority to two of the community councils, one close to the city, one in a suburban area of the municipality. The initiative for the CC experiment came from the Conservative Party holding the mayoral position at this time (1985-86). The experiment was launched as a means of breaking with a Social Democratic past. The Conservative Party used terms like "real local autonomy", "freedom from the straitjacket of the state", "debureaucratization" etc. (Lie & Hauge 1991: 15-16). By the of the experiment with CCs was implemented the context for the CCs had changed in two ways. First, there had been a political shift so that the mayor was now a Social Democrat. Second, the municipality had experienced a serious economic decline. Since the decision to establish CCs was based on a broad political consensus, the change in political power didn't affect the process in any significant way. The economic recession, however, meant that the administration opposed the idea, and questioned whether the municipality could afford it. The result was a decision to establish CCs increasing municipal expenditures (Lie & Hauge, 1991: 32-33). The conflict between politicians and administration caused some problems in the beginning, but it is now reported that the situation has improved (Olsen & Rommetvedt, 1993). Likewise, the economic problems have posed certain limits on the CCs, but have not stopped the process.

The experiment contains two forms of decentralization, both a political and an administrative one. The two CCs were given extended political discretion supported by the establishment of local city administrations. The administration at the community level is organized at a 'cross-sectorial' basis. As the only one in our case material, the case of Stavanger thus concerns both efforts for more efficient provision of public service by a cross-sectorial organized administration and of a more democratic process of decision making through the CCs.

The members of the CC are appointed by the local council, and not by direct election. The recruitment of the CC is partisan, the political parties are given seats proportional to their strength in the municipal council. The CCs have 11 permanent members. The size of the population in the two districts covered by CCs are approx. 10.000 in both.

The community councils have achieved a considerable scope of discretion in the local decision-making process within most of the public service sectors in the area. This includes primary schools, cultural policies, child-care, care for the elderly, welfare services and payments, public housing and so on. The experiment also includes the establishment of an administrative staff. The local administration is also the executive secretary of the CC, with a managing director on top.

To what extent do the experimental councils satisfy our previously mentioned criterion of democratization?

The CC administration has, to some extent, institutionalized a local public sphere. The CC meetings, however, do not differ very much from meetings in other political committees, and are of limited interest to the individual citizen. Normally s/he attends the meetings only when s/he has a special interest.

What about the third criterion, the CC as a channel of bottom up influence? For the citizens, the CC at least represents a channel of information. A questionnaire to the members of the CC and samples of the population concludes that both the politicians and the local citizens are more concerned with local issues than they were before (Olsen & Rommetvedt 1993: 99). The experiment has helped to increase the citizens level of knowledge in such a way that their ability to influence the decision-making process has increased. For the members of the CCs a learning effect might be a consequence too.

So far there has been several difficulties regarding mobilization of the population. It has been proven difficult to activate people through more intimate interaction with the CC. However, there are positive results especially in narrowing the distance to the authorities both as users and as voters in relation to the politicians.

The role of the CCs as a recruitment channel is hard to estimate at this moment. Political careers starting in the CC and ending up on some other level of authority are not discernible at this point. The evaluators have noticed that the members of the CC state that participation has been more interesting than before as a consequence of the CC`s new role. They also say that the politics that arises from the political parties are less relevant at the community level. This can indicate less interest in participating in politics at the municipal level.

Several evaluation reports have documented the merits and effects of the CC project. Many of them, especially those carried out by "Rogalandsforskning", conclude that the CC experiment has a clear and specific direction towards democratization (Lie & Hauge, 1991, Olsen & Rommetvedt 1993). Among the results mentioned are the simplification and shortening of the decision-making process. The councillors report that participation in the CC is more meaningful and that the CCs have been understood to play a far more distinct and offensive role than before they were granted formal authority (Lie & Hauge, 1991: 31).

2.2. Tromsø (Norway)

The municipality of Tromsø has both an urban and a rural character. Within the municipality is the city of Tromsø, surrounded by several rural communities, most of them very small, some without road access to the centre. The rural aspect of Tromsø gives a special character to the city. To face this two-fold settlement structure, the municipal authority of Tromsø in 1986 invited 16 rural communities to establish "development committees" (: "utviklingslag"). The rural area of Tromsø covers an area of diversity both with regard to scale and demographic structure. The population of the settlements varies from about 20 to nearly 1000 inhabitants. The average size is about 300.

The idea to take this sort of action was motivated by the fact that these areas had several problems. For example, a diminishing and aging population, an increase in out-migration, and an under developed business structure.

These organizations did not, at the beginning, gain any formal authority in the area. They were supposed to be informal groups working in close cooperation with the local population, as well as in cooperation with the municipal authorities. The goal, as the municipality put it, was to develop local services, the local business structure and the environment in general, simply by activating local resources. The philosophy was to stimulate the people's own competence and ability to solve community problems. The communities were given an amount of 20.000 Nkr (approx. $3000) each if they established "development committees" by popular election.

Today, these organizations still exist, and have even increased in numbers. At this moment there are 22 development committees. Their authority has, to some degree, increased somewhere along the road at least in some sectors. For example, they have obtained the right to influence questions concerning land use and planning. They are also heard in other questions of interest to the community. Their real influence, however, mainly rests in informal structures (except for the 20.000 Nkr. a year they are given for free disposal), in the way they solve problems, in the intimate cooperation with the authorities and in handling local matters in their own way. The organizations also initiate matters and ideas and present them to the authorities. The bottom-up structures are well established and very often achieve results.

The committees of these organizations are elected directly by the inhabitants at the annual meeting. The whole adult population in the area has the right to participate and to vote. The number of committee members varies from 5 to 9. The members are not elected on the basis of a political party manifestos, but only as representatives of the local citizens.

Over the years, the CC has been engaged in a large number of issues. What is on the agenda at a specific moment is always a question of how the local situation is defined by the participants. Most of the local organizations have been involved in local land-use planning, the deployment of road-lightening-systems, welfare services for old people, in taking actions to secure communications in the area , local culture and leisure-time activities, stimulating the business structure and so on. It is, therefore, possible to treat "development committees" as community councils proper.

In what way do these local organizations satisfy our criteria for increased democratization?

The CC`s have managed to mobilize the population in the area to a great extent. First of all, this is documented by the expansions to new areas, and the continuity of the organizations once established. The participation, however, is more than the CC meetings. Most of the CC's have established permanent "working-committees" on several subjects concerning the community. In this way, more than just the CC members participate in the day to day activities of the CC.

However, the participation of the citizens in the annual meetings has decreased over the years, as well as the number of working-committees. For the first few years more or less the whole population of the community participated in the meetings. Now just those who have a special interest attend the annual meetings. This may be troublesome, but it is also "natural". Even if the activity seems to slow down a bit, the CC seems able to mobilize the citizens whenever needed. Periods of lower activity may be interpreted more as a "rest between the battles", more than as a fading out of the CC.

The CC, with its working committees, definitely represents a local public sphere that was not there before. The debates on local issues, carried out at committee or annual meetings are stronger, more specific and better articulated than before. The CCs have also managed to integrate different local voluntary and ownership-based interests in the committees, partly due to instructions from the municipal authorities. Local conflicts are handled within this context, not suppressed, something that otherwise might have caused problems like particularisation and segmentation of local interests and local debates.

The function as a channel of recruitment to other levels of participation is weak, but not totally absent. A low share (20%) of the committee members are also members of a political party. This percentage is just slightly above what is the average participation rate in Norway. There is a very weak link between the local organisations and the parties. However, a small number state that they have been more motivated to run for election because of their participation in the CC. So, perhaps there is some vitalizing effects on democracy in this sort of participation.

What about influence? To this question it is impossible to give an exact answer. The degree of influence varies from issue to issue and from one community to another. Some committees are more astute than others in their actions, and in finding ways of pushing through their opinions, plans and operations. Some committees are definitely operating on the municipal backstage.

A special effort has been made to stimulate the different CCs to cooperate. A meeting for the leaders of every local committee has been instituted four times a year. This constrains a potential development of narrowness (or NIMBY-effect) in the small communities and stimulates the identities of being a part of the rural area of the municipality, as well as cooperative actions towards local or central authorities. However, these meetings are a top-down municipal authority initiative, and so is the organization of the meetings.

A special remark must be made at this point. The municipal authority, especially the staff of the chief administrative officer, has acted very sensibly and competently in the matters discussed here. Their willingness to cooperate with the local CCs, the resources that are made available for the

CCs, (not only the amount of money transferred once a year, but also resources like personal, specially trained secretaries, and special concern with the rural areas in most of the local government policy,) shows that the policy towards the rural areas is well integrated both in formal and in informal structures in the political system, but mostly in the administrative culture of the local authority.

2.3. Herlev (Denmark)

Herlev is an urban municipality within the Copenhagen Capital region. Since 1986, two of the districts have had a CC. The CC areas are comprised of between 4000 and 7000 inhabitants, and covers about half of the population of the municipality. In January 1995, the two CC`s were expanded to four, which meant that every district in the municipality now has their own CC. The CCs have 9 permanent members, who are appointed by the local authority in the same way as other committees, i.e. composed by political party members. The members also have to be inhabitants of the area. In addition to the 9 permanent members, the public service institutions in the area are represented in the meetings with 4 observers altogether. The observers do not have the right to vote.

The CC enjoys a high degree of authority and makes decisions in several areas, ie. school policy, leisure activities, social care, environmental issues, local planning, and so on. The CCs have a budget of about 10 mill DKr (approx. $1,4 mill) each. Within this budget there are possibilities for free disposals, at least in some of the items. However this freedom is limited since most of the items in the budget are fixed costs, like salaries, and can not be changed. The CCs are also free to take up any issue that may be of concern to the municipality.

The formal structures of the CC are more or less the same as that of the municipal authority, apart from the presence of the observers. One main difference, however, is that anyone in the community can attend a CC meeting, and they are free to participate in the debate before decisions are made. This is a principle of open meetings. This is not the case in the municipal council. Even though meetings are open to citizens they can not speak without explicit permission. One of the CCs does not arrange closed meetings before the open part to avoid closing the decision-making process too much. The willingness to make compromises is strong. The political parties are not very prominent or visible in the deliberations of the CCs. Questions and issues are seldom put to the vote.

A special feature of the Herlev experiment is that the CC`s interaction is mainly directed towards the local public service institutions, local voluntary organizations, and associations, not the citizens directly. Most of the initiatives taken by the CC are matters of interest mainly to these institutions, e.g. "green days", "the days of the associations" and so on.

How do we evaluate this case regarding the effects on democracy? The Herlev case shows the CCs enjoying a high degree of formal and real authority. This means that local autonomy is to a large extent achieved. To some degree the election of the CC member is a problem, due to the fact that they are elected indirectly. Although the members have to live in the area, this is not enough to secure local representation. Since they have not been elected by the citizens in the area, their local legitimacy is weak.

If we turn to the question of mobilizing the citizens in general, this might be the central problem of the CC in Herlev. Cooperation with the institutions and organizations are successful when it comes to questions of common concern. However, the citizens are not intervening into the affairs of the CC.

The CCs have managed to strengthen the local public sphere. The local newspaper covers all meetings, makes reports of different issues and so on. But perhaps the form of the meetings is even more important. Open meetings and debates have put the decision making process into the public sphere much more than the municipal authority has managed to do. First of all the CC functions as local public spheres to the voluntary organisations and institutions in the area. They have their own observers at the meetings, participate in the debates, and influence the agenda in several ways. The local public sphere is integrated in the institutions, but not in the local community as such. Usually, the individual citizen does not attend the meetings. Personal involvement in one of the cases seems a necessary prerequisite.

When it comes to the ability of influencing the local agenda, the CCs have had effects. The members and observers are "locals", which means that they are sensitive to local questions and suggestions. The CCs in Herlev have a high degree of authority and a large budget, which altogether represents real autonomy. However, one should be careful to consider large budgets and autonomy as equals. The budgets are to a large degree fixed, by wages, maintenance, etc.

To the question regarding the CC as channels of recruitment to higher levels of participation, our data give us no clear answer. Generally one may remark that because the members of the CCs are selected from the political parties, the CC`s possibility to function as a new channel of recruitment is reduced as the councillors are already party members. The

CC system may be productive as a training ground for party members before entering into municipal arenas.

To summarize, our preliminary conclusions on the Herlev case is this: Since the specific gains are relatively few, the CCs have mainly created an opportunity for increased democracy, and only to some degree have actually realised this opportunity.

2. 4. Ejby (Denmark)

Ejby is a rural municipality on the island of Fyn, in the middle of Denmark. The centre, Ejby, is surrounded by small countryside villages. 5 of these were separate municipalities before the amalgamation in 1970. The total population in the municipality is approx. 7000 persons, of which 1500 live in the centre, Ejby. In both economic, cultural and social terms, the municipality may be characterized as a traditional one. The communities in Ejby are primarily agrarian, but in the centre and in a nearby village, Gelsted, there are some industries. When the idea of the CCs were formulated, the municipality experienced some problems with the local development: Reduction in local employment, the closing of local shops, reduction in the recruitment to the local schools and so on. The process leading to the establishment of CCs started in 1982. The municipality administration invited the populations in the local communities to present ideas on how to handle the economic crisis and develop the municipality. The ideas centred around the topic of decentralization and self government. The idea of the CCs were designed in cooperation with the administration and the Municipality Council (1989). Some months later, CCs were established in three of the surrounding villages, with the CC members elected at public meetings. Later, in 1990, the elections were formally confirmed by the local government. Formally the CCs are appointed by the local government. However, in Ejby the procedure is simply that the municipality council 'appoints' those already elected by the local community. Two more CCs were established in 1991, and the last one, in Ejby, in 1992. So there are now 6 CCs in the municipality of Ejby, covering the entire area. On average the CCs cover between 500 and 1500 inhabitants.

The CCs in the Ejby case may be characterized by diversity. What they have in common is the small amount of money from the local authority (10.000 Dkr per year (approx. $1.400)) and the absence of authority other than the right to advise the municipal council on local issues. The members of the CC are elected directly by the citizens, by secret vote or at open meetings. Candidates are not nominated by the political parties. Most candidates are elected at open meetings, but in one CC they are put to a vote by a specific ballot during the local government elections. Some CCs have formalized representations from local associations and institutions. One of the CC's limits their activities to the administration of the "community house".

Even in the beginning if the Ejby case presupposed a transformation of authority to the CC, this never happened. The CCs have had no fixed mandates,. nor have they been granted discretionary competence. The activities of the CCs have been characterized by issues like raising matters, case by case, not as a totality, through formal or informal channels of influence. The CCs have raised a wide range of issues. One CC has managed to establish a child care institution, with the help of volunteer parents. Another has arranged a system of transporting old people to the library. A third one has organized demonstrations and petitions against plans for changing the main road system through the village.

Has the establishment of CCs meant a strengthened local democracy? Regarding our first criterion, we say that a local public sphere is established. Most of the CCs have had regular meetings, and the meetings are open to the public. However, the citizens usually do not attend the meetings. One of the CCs has on several occasions managed to mobilize the citizens on specific issues. In one occasion, in Gelsted, more than 100 persons attended an open meeting concerning the future of the local village. A local newspaper covers most of the activities of the CCs, and we have good reason to believe that people monitor the CC activities in this way.

Second, when we look at the CC`s ability to mobilize in general there is no doubt that Ejby has been successful. However, the ability to mobilize is concentrated within some specific problems. When the sources of local issues are drying up, so does the people's interest. The CC's also represent an increase in the local publicity. Several of the communities have established 'community newsletters' or revived old ones. Open meetings are also increasing publicity. The question, however, is if the experiment is fading out. Time will tell. As mentioned before, the municipal agenda is now increasingly responding to local needs and interests, and several issues have been raised directly as a result of the CC activities. Of course this does not mean that the CCs have been successful everywhere (the protests against changing the main road was not successful). The formal authority of the CC is still limited.

There have been some problems in the CC`s ability to recruit councillors. Three reasons are mentioned. The first reason is that membership is a massive burden. The CC memberships are often added to participation in several other community committees. The second reason is some sort of disappointment about the CC`s lack of authority. Because of this, the newly established user-committees established to administer the municipal service institutions in the local community are seen as competitors, and the CC looses members to these committees which have more influence. The third reason is the success of the CC itself. This might look like a paradox. However, some of the CCs have accomplished what they wanted, they have their own school or whatever. Now they have no new ideas and the motivation seems to be waning.

Regarding bottom up influence, the effect is limited by the lack of autonomy. Even though the Municipality Council takes the wants of the CCs seriously, we would like to point out that whatever small increase in autonomy the CCs may have achieved, most nevertheless remains in the hands of the municipal council. Formally it is the municipal authority that appoints the members of the CC, and it is the municipal council which has the final word in all major decisions on matters of community interest.

When it comes to the CCs as a channel of recruitment, we have been informed about one case where one CC leader ran for election to the local government. Whether this is a sign of a broader tendency, we do not know at this moment.

To summarize, the Ejby experiment scores on most of the criteria. We would especially like to point to the way the members are elected, and their continuous efforts of being responsive to community wants and interests. The members obtain their political legitimacy directly from the community. The lack of formal authority and a very limited budget however, as well as the consequence of less formalized and institutionalized structures, may lead to a situation where the gains disappear.

3. Four Cases - A Comparative Perspective

In summary, what can we say about the four cases? Are there any marked differences and patterns? As mentioned above, we operate with factors of explanation on two different levels: the national and the municipal level[2].

[2]Of course differences are also found in the functioning of the CCs within the single municipalities. We do not consider such differences in this paper, but we would like to suggest some principles of explanation for sub-municipality differences. Three different modes of explanation seem possible. The first (and perhaps least interesting for the social sciences) is the so called "personal factor": different people do the job differently, and especially the chairman-position seems to be important for CC functioning. The second is different traditions in the CCs themselves. When the institutions have worked for some time, specific cultures of

Unfortunately the two institutional dimensions on the municipal level, indirect appointment and a high degree of decision power co-vary[3]. The main problem for this project is however, that these lines of differences also follows the urban - rural dimension. In the rural settings the CCs have small budgets and no (or very little) formalized power, and in the urban contexts the CCs have a high degree of formal autonomy.

Below we present our findings in a matrix, trying to clarify the answers given in the four cases.

Principles of democracy / Municipality	Institutionalization of a public sphere	Mobilization	Agenda setting	Channel of recruitment
Tromsø (No)	Yes	High degree in the beginning, now lower	Yes	A small number
Stavanger (No)	To some extent	Several difficulties	So far only potentially through increase in citizens knowledge on local affairs	None reported
Ejby (Dk)	Yes	Succesful towards single issues in the beginning. Now declining	Yes	Very limited
Herlev (Dk)	A certain strengthening through the openness of the meetings	Low degree	Yes, but in relation to institutions and associations more than citizens	Very limited, but some amongst the 'observers'

Concerning the first three of the criteria (public sphere, mobilization and agenda setting) there is a close correspondence of performance between the CCs in the two rural municipalities (Tromsø and Ejby) and between the CCs in the two urban (Stavanger and Herlev) municipalities. There does not seem to be any national patterns of difference in our cases.

Generally success in terms of democratization seems to be greatest in Tromsø and Ejby. Concerning both agenda setting and institutionalization

decision making emerge. The third, that might be called 'local contingencies', refers to the local community as a particular social setting different from other settings. It is obviously insufficient only to distinguish between urban and rural, since these categories cover a vast number of highly different social realities.

[3]This stems from the fact that it is a legally prescribed demand that if the CCs are to have real decision power and influence on the budgets, their members must be appointed by the local government. However, the Norwegian Government is now considering an experiment with a combination of direct elections and formalized power.

of a public sphere the CCs seem to have succeeded. In Stavanger and Herlev a local public sphere has been institutionalized as well, but with a far more limited scope. First of all, the CCs function as local public spheres to the voluntary organisations and institutions in the area. The local public sphere can be said to be integrated into the institutions, but not into the local community as such.

In terms of mobilization the pattern is the same. There is none or very little or no national variation,only a clear line of difference between the rural and the urban CCs. Again, it is in the rural CCs that the greatest mobilisation has been found. It is worth mentioning that this picture seems to change over time, so that the CCs in both the rural settings have experienced a decline in active citizens involvement. But even after a decline, the relative number of actively participating citizens is significantly higher than in the two urban municipalities.

When it comes to the fourth criteria, the effect of the CC on recruitment to municipality councils , the CCs in the cases studied have had only very limited, but still slightly positive results.

We shall now give some possible explanations for these findings, in terms of factors at the national and at the municipal levels.

3.1. National Differences Between Norway and Denmark

The differences in the functioning of the CCs do not follow national lines. This is relatively easily explained, since the relevant national differences between Norway and Denmark are very small. In terms of political culture both countries share a Scandinavian tradition of welfare with a strong egalitarianism and relatively strong state interventions, collective responsibility etc. In terms of institutional arrangements the differences are small as well. However, the overall structure of municipalities is somewhat different. Both Denmark and Norway have carried out a municipal amalgamation in the 1960s but the decrease in numbers was far greater in Denmark than in Norway (Gunnel Gustafsson, 1980: 340). But in our case material the difference between Denmark and Norway is not shown since the present size of the municipalities of Ejby and Tromsø are both results of amalgamations.

Second, the reform strategies aimed at the local government level differ. In Denmark there has been a reform of decentralization of power from the (territorially defined) local government to the (functionally defined) public service institutions, and establishment of governing bodies with user participation of the institution, so called 'user-democracy'. In Norway the principle of citizen- (as different from 'user') participation is underlined in reform proposals. This does not explain the function of the CCs, but it might explain why CCs are a more common phenomena in Norway than in Denmark [4]. The legislation on CCs is, however, very much alike in the two countries. Any municipality can establish CCs, but it is a condition of any delegation of formal authority that their members must be appointed by the local government. Direct elections are only possible if the CC does not enjoy formal authority.[5]

There might be other types of differences between the two countries, but not along the criteria we have employed in this chapter. So we have to search for other factors of explanation that might shed some light on the differences found. Therefore we shall move to the municipality level. At that level two different explanation factors are given, the institutional arrangements and the social context in terms of the distinction between urban and rural settings. Unfortunately the two institutional dimensions on the municipal level, indirect appointment and a high degree of decision power co-vary. The main problem for this study is however, that these lines of difference also follows the urban - rural dimension. In the rural settings, the CCs have small budgets and no (or very little) formalized power, and in the urban contexts the CCs have a high degree of formal autonomy.

This of course means that there is principally no way of determining whether it is the institutional arrangements or the social settings that may explain the differences found. Of course there is no necessary contradiction between the two types, they might also reinforce each other, but no final conclusion can be reached on the basis of our material. Therefore, we shall go through both types of factors and discuss how these might explain our findings.

3.2. Institutional Arrangements

In what way can the institutional arrangement help to explain the higher degree of success in Ejby and Tromsø than in Stavanger and Herlev? We have three basic differences in institutional forms. The first is the question

[4] In Sweden there has been two tendencies. In the 1980s there were many experiments on decentralization to CCs. But in the '90s (after the economic recession) the tendency is much more in the direction of 'privatisation' and market. We shall return to the Swedish experiences with CCs later in this article.

[5] However, the Norwegian Government is now considering an experiment with a combination of direct elections and formalized power.

of the appointment method. The second is the degree of formalized power, or authority in decision making. The third is financial resources[6]. These institutional arrangements are by no doubt very important for understanding the functioning of the CCs, and the municipalities in each country may choose between different forms.

Regarding the first factor, there seems to be a very direct relationship between principles of appointment and degree of success for the CCs. In Herlev and Stavanger the CC boards are appointed by the municipality council. In Tromsø and Ejby the members are chosen through direct elections by the local community. Here it seems - as would be expected - that direct elections lead to a higher degree of active citizens involvement. Direct election both attracts attention towards the CCs and strengthens their legitimacy, each of which increases the inclination towards active participation.

When it comes to the second and the third factor - formalised authority and financial resources - the picture is more ambiguous. On the one hand one would expect that a high degree of local autonomy - as in Herlev and Stavanger - would lead to more interest in the CCs' affairs on part of the citizens and vice versa. This is, however, not the case. On the other hand the hypothesis can be made that a high degree of formalization leads to bureaucratization and alienation of the citizens towards the CCs. Or it could be argued that when the CCs enjoy only a small amount of formal competence, they have to confront the centre (normally the municipality council) with issues of public concern. This seems to stimulate both a local debate and local mobilization. Most examples of a high degree of mobilization and debate are found when there are conflicts or disagreements between the community and the local authorities in a specific matter, something that is less likely to happen when the areas of local autonomy are clearly specified. It remains an open question, whether this second explanation is the best or it is simply a question of the positive effects of the first being outweighed by other negative effects.

---

[6]In our cases formalised authority goes together with a high degree of financial autonomy on the part of the CCs. These two factors will therefore be treated together, but it should be mentioned that such a connection is not a necessity. It is easy to imagine a situation where CCs enjoy a high degree of formal authority but lack financial support. In such a case the real influence would of course be very limited.

3.3. Rural/Urban Differences

The third line for explanation of the variation in degree of CC success is the rural - urban distinction. Is there something in rural settings that, in contrast to urban settings supports initiatives such as CCs? The hypothesis is that in a rural setting there is an already existing feeling of "common spirit", that makes broad participation relatively easily achieved. Such a feeling of common spirit is supported by relations such as relative proximity in the rural community, longer historical roots and a relatively smaller degree of mobility, that the community is relatively well geographically delimited, and also the simple fact of numerically smaller populations.

This could also be stated in terms of a political community that tends to appear more spontaneous and to follow the lines of the local community as such, which gives a generally higher degree in citizens' interest in common concerns of the local community. Also the 'common good' of the community will be relatively easily defined and uncontested, and often placed in opposition to 'the central authority' i.e. the local government. This also supports citizens' engagement in the affairs of the community. In urban (and modern) settings the inhabitants will be less immediately directed towards the locality. This is due to a higher degree of mobility, the unclear limits of the local community, the possibility to satisfy wants in many places other than the local community and the sheer size. This means that the political community will be much more of a problem, something to be constructed more than taken for granted. And it will not necessarily correspond to the limits of the defined local community, if the social context can be conceptualized as *a community*. Therefore the inclination to participate will be weaker than in the case of the rural communities. Likewise, the common good will be much more contested and diffuse, and as a consequence more difficult to place in opposition to the centre, with a lack of 'mobilization-potential' as the result.

In sum, it does seem that the social settings have an impact on the functioning of political institutions as CCs.

4. Examples from Sweden

The Nordic perspective is not complete without examples from Sweden too. In Sweden, the CC was institutionalized in 1980 with "The submunicipal reform", a specific law permitting local authorities to establish CC, or "kommunedelsnämnder" in different parts of the territory. The "fourth level" of government was institutionalized. Presented in the early 1980`s, this reform was believed to be a very important instrument indeed to enliven local democracy and adjust the services to local needs.

Until 1991 the CC had been introduced in 37 municipalities, whereas 16 were totally covered by CCs (Gustafsson 1992). In the rest of the municipalities, CCs were introduced in some parts and not in others. The variations in structure has increased in the last couple of years. There is not one dominating model. The amount of money controlled by the CCs varies a great deal between the municipalities, from "pocket money" in Umeå to 2/3 of the total budget of the municipality in Örebro.

Speaking in institutional terms, they do have some characteristics in common. They are not allowed to set their own taxes, the members are elected indirectly and the committee meetings are mostly closed. All of them however have some competence of decision making and authority in these matters. However, the number of such highly competent CC's has decreased in the last couple of years.

If we ask the same questions the Swedish case as to the four cases we have examined in this article, as positive results can be recognized;

- Participation is slightly improved. The number of representative posts has increased and more people are given political responsibilities. Also people who have not previously been engaged in municipal affairs are assigned such positions. Citizens participate in official meetings when the issues in question are of importance to them. Even "passive" citizens have become engaged, although to a very limited degree.
- The selection of representatives functions more or less the same way regardless of neighbourhood councils. Some new recruitment is recognised.
- The opportunity for citizens to articulate their demands and wants have been improved. They obtain yet another channel of influence.
- The conditions for communication between elector and elected are changed in several ways by the introduction of neighbourhood councils.

However, the expectations and hopes of reduction in bureaucracy, quicker decisions, shorter time from policy initiative to implementation and integration of sectors and professions have been difficult to fulfill. Furthermore, citizen participation and influence slightly improved and small increase in party activity can be demonstrated. (Kolam, 1987, Gustafsson 1992, Nilsson 1993, Montin 1989, 1993).

The expected large increase in citizen involvement has failed to materialize. Few have participated in open committee meetings and information meetings or made contact with sub-municipal politicians. Many citizens state that they are not aware that such activities take place in their area. Obviously they cannot, therefore, use the neighbourhood council as a mediator for their needs and demands. Citizen initiatives are collective rather than private. Local organisations are rather active towards the community councils, the single citizen is not. Several studies indicate that it is the citizens who are already active who have obtained yet another channel through which they can express their points of view and possibly exert influence (Kolam 1987). It is also possible to trace some growth in bureaucracy. In many places CCs have become yet another step in the decision making process.

5. Concluding Remarks

The main purpose of this chapter has been to give an answer to the question of the CCs effect on the local democracy.

To the question "direct or representative democracy" the answer seems to be the latter. In our case-material we have examples of communities with a population as small as 70 - 80 persons (Tromsø). Even in such reduced social spaces the CCs report a relatively weak citizens participation in the day to day activities. So it is plausible to conclude that institutional reform in itself does not change the representative democracy into a direct one. However it would not be correct simply to interpret the material as a victory for representative democracy as such. Instead we argue that CCs are best seen as a change in the functioning of representative democracy, where elements traditionally connected with direct democracy are incorporated into a basically representative system.

We would mention 3 such changes. The first is the open board meetings, where everybody is allowed to speak and argue for his or her case, before a decision is made. This is an obvious difference from the local governments, where non-members are allowed only to pose questions before the meeting and only on matters not on the agenda. The second is the recruitment of a different sort of politician, who presents himself as representatives less of specific interests (in contrast to politicians at central levels) than of the 'common good' for the local community. This would also make them more responsive to the citizens' point of view between elections. This tendency is of course strongest where the members are elected directly (and often on an explicit non-party basis) than where they are appointed. However, it is a common trend that the members of the CCs appear to be more responsive than politicians at more central levels. Thirdly, we would point to the use of "citizens-meetings" as a way of integrating elements from direct democracy into the representative system. Such meetings are of course primarily found where the members are elected directly. In these CCs the annual citizens meetings have a high degree of direct impact on the agenda of the CCs, and in certain cases also on the appointment to the different positions in the CCs.

The second concluding point is whether the CCs improve democracy at the community level at the price of a lower overall participation at the municipality level, or to put it differently, whether local democracy is a kind of zero sum game. Even though there only seem to be little gained in terms of recruitment from CCs to higher levels, the opposite does not seem to be the case. Nowhere was it reported that it had become a bigger problem to recruit members to the municipal council after the introduction of CCs. And for those CC members who did not want to run for seats in local government the alternative to CCs seemed to be no activity. Likewise, those who reported an interest in local government did not see the CC as an exclusive alternative, but rather 'a stop along the road'. So to the extent that they have been a real gain for local democracy, the relationship between CCs and the central local government should not be seen as a zero sum- but as a plus sum game.

The decisive question is whether CCs are a real plus for local democracy. We contend that they are, but the picture is ambiguous. The gains in terms of institutionalizing a public sphere and in bringing the agenda of the community more in correspondence with the needs of the community itself, seem to be obvious in the two rural and less formalized cases (Tromsø and Ejby), but only small or even only potential in the urban and highly formalized cases (Stavanger and Herlev). The mobilization of the citizens seems to follow the same lines, and in the two rural cases a tendency towards a decline in mobilization can be traced.

Finally, we would like to point to two democracy dilemmas that could be read from the results. The first is the possible dilemma between formalized power and citizen mobilization. If real democracy is dependent on real influence, there might be a dilemma here, since it seems that formalization of influence carries the price of less mobilization. One might put it this way: in order to secure real citizens control of decisions there must be certain channels for influence and articulation of disagreement towards higher levels. So some kind of formalized institutionalization is necessary. However 'too much' institutionalization might have the effect of reducing citizens mobilization.

The second problem regards the question of the relation between the political community and the surrounding social community. On the basis of our findings it could be argued that it is easier to establish a public sphere or a political community in settings where these tend to follow the lines of the surrounding community (i.e. in rural or more 'traditional' social contexts). So with the overall tendencies towards less homogeneous social settings one could argue that a major problem for future democracy is how to create and maintain political communities, on the basis of democracy itself more than on given social communities.

Despite the fact that they are not unequivocal successes, CCs do not seem to be the worst answer to some democracy problems.

REFERENCES

Andersson, M. et al. 1989. *Lokaludvalg - en vej til demokrati og effektivitet?* Copenhagen: AKF`s forlag.

Andersen, J. et al. 1993. *Medborgerskab - Demokrati og politisk deltagelse.* Herning: Systime.

Berg, Trond (ed), 1983. *Deltakerdemokratiet.* Oslo: Universitetsforlaget.

Carlsson, Y. 1992. *Det kompliserte nærmiljøarbeidet,* NIBR-rapport No. 12, Norsk Institutt for By- og regionforskning.

Gustafsson, Agne 1992. "Problem och möjligheter vid inomkommunal decentralisering". *Nordisk Administrativt Tidskrift* 2/1992.

Gustafsson, Gunnel 1980. "Modes and Effects of Local Government Mergers in Scandinavia". *West European Politics,* Vol 3, No.3.

Held, D. 1990 (1987). *Models of Democracy.* Oxford: Polity Press.

Koch, H. 1981 (1945). *Hvad er demokrati?* Copenhagen: Gyldendal.

Kolam, Kerstin (1987). *Lokala organ i Norden 1968-1986. Från ide till verklighet.* Umeå: 1987.

Lie, T, & J. Hauge. 1991. *Omstilling med bremsene på. Evaluering av frikommuneforsøka med bydelsforvaltning i Stavanger og Trondheim.* Rogalandsforskning, rapport nr 31.

Montin, Stig. 1989. "Från demokrati till management. Decentralisering inom kommunerna". *Statsvetenskaplig Tidskrift.* No.2, 1989.

Montin, Stig. 1993. *Swedish Local Government in Transition.* Örebro Studies 8. (Högskolan i Örebro).

Olsen, K. H. & H. Rommetvedt. 1993. *Desentralisert bydelsforvaltning - fleksibel likebehandling og demokratisk effektivitet?*. Rogalandsforskning. Rapport nr 99.

Rose, L. (ed.) 1991. *Det er lov å prøve seg. Forsøk, reorganisering og ledelse i kommunene*. Oslo, kommuneforlaget.

Pateman, C. 1989 (1970). *Particitation and Democratic Theory*. Cambridge Univ. Press.

Ross, A. 1967 (1946) *Hvorfor demokrati?* Copenhagen: Nyt Nordisk forlag. Arnold Busch.

Togeby, L. 1992. "The Nature of Declining Party Membership in Denmark: Causes and Consequenses" in *Scandinavian Political Studies* Vol 15 No 1

Chapter 7

"Learning User Influence" the Constitution of User-Subjects in a Differentiated Welfare State[1]

Hans Wadskjær
Aalborg University, Denmark

Introduction

In the modern crisis-ridden context of the welfare state I argue in favour of a decentralised tripartite model connecting the local political level, the mediating professional level and the users. The size of the budget is seen, primarily, as being a constraint. In a case study concerning elderly care, the constitution of the model will be demonstrated. The creation of a user subject is shown to be necessary in order to circumvent double false consensus in a supportive setting. In the discussion I, furthermore, include primary schools to illustrate the existence of insufficient conditions within the system for the development of user influence.

In this text I use the terms user and user forum as empirical terms, whereas the concepts of user role and user subject are related to the con-

[1] I would like to thank research fellow Catharina Juul Kristensen for inspiring discussions whilst translating the article.

-cept of consciousness, and in the case of the latter, additionally to civil society (Kean, 1988). The User Subject is conceived of as a collectively reflective actor, differentiated out and (re)integrated as a part of a transformed model of the Welfare State. Moreover, when conceptualized as the **user** subject, it is as such confined, i.e. in a differentiated model of the Welfare State, to act through influence as a (Parsonian) medium (Parsons, 1967). As a real type the user subject has to be constituted as a manifestation of civil society, and it has to be aware of itself as an actor. As I refer from the outset to a 'grounded theory' approach (Glaser & Strauss, 1967), I primarily apply a real type version of the user subject.

Reproducing the Welfare State

Most of the North European welfare states are drifting towards market or quasi-market solutions. The users are being offered a greater choice of welfare services. In the case which is reported in this article, the users are offered the opportunity to influence the quality of services. The possibility of a non-market route towards welfare decentralisation is demonstrated by means of the users acting as a manifest expression of civil society, i.e. as user subjects.

As Hirschman has argued, the involvement of citizens may shift between collective political and individual market action preferences, as a result of individual and collective learning processes. The experience of disappointment plays an important role in Hirschman's analysis of shifting involvement in the public sphere. On the crisis of the Welfare State, he advances another perspective than that of fundamental contradictions;

> "Rather, it sees these difficulties as serious, but quite possibly temporary growing pains. Those pains .... can eventually be brought under control as a result of various learning experience and mutual adjustments" (Hirschman, 1982, p.43).

The shifting involvement, which is caused by changes in individual motivation, corresponds on the societal level to a changing emphasis on the steering media, power and money (Parsons, 1967). Implicitly Hirschman suggests a relationship between structure and action, with emphasis on the bottom up aspect. In so doing, he also suggests what could be called an active reproductive aspect for the development of society and its institutions, including street level welfare providing institutions (Hirschman,

1970, 1982). The expansion of the Welfare State can be said to postpone the moment of truth, i.e. the crisis of the Welfare State.

In recent decades the reproduction of welfare in Scandinavia has run into problems. The slowing down of economic growth combined with large public budgetary deficits and unemployment high rates are significant constraints on the reproduction of universal tax-based welfare.

Reproducing the Welfare State, to say nothing about reducing it, is quite a different thing than to create or expand it. The welfare arrangement has become a structural fact of society, implying that the entire project may have lost its dynamic perspective. It may have exhausted what Habermas has called the utopian energies (Habermas, 1990), a point which is compatible with Hirschman's notion of disappointment.

The Mediating Link

The arrival of the Welfare State ensued the establishment of a considerable mediating link. The Welfare Project became a ladder for the professions; the doctors, the teachers, the nurses, and to a lesser extent, groups such as the social workers. The professional associations, the White Collar Unions, and other groups working within the Welfare State expanded their power base and succeeded in obtaining and preserving a substantial amount of autonomy. While the Public Choice School has theorized one-sidedly on the self-interest of professions and bureaucrats, I draw attention to the mediating link as a multitude of subjects, which in the Welfare State context, can be expected to act reflexively. The professions still have their projects. They participate in the formation of welfare policy through the corporate channels, and implement policies under conditions negotiated between the Trade Unions and Government.

The Modern Condition

Equally important for the development of the Welfare State, we are witnessing a deepening of the Modern Condition in Society, a condition Giddens termed "high modernity" (Giddens, 1991). According to Giddens, one of the aspects of High Modernity is the development of a new self-identity. Everyday life has become more reflexive. The modern conditions have pushed Man in the direction of a self-referential entity. People are, as subjects, on the way to becoming their own individual project. Beck and Giddens have analyzed society as a Risk Society, implying that the individual person is able to influence his/her life chance by making, for them, advantageous choices. In other words people aim at colonizing their future.

The advent of people with this new self-identity, may shift the focus from disappointment in the Welfare Service, to the reflexive use, and effort to influence them, and hence a new motivation for involvement.

Another significant aspect of High Modernity is the role of the sciences; the critical energy of Science has turned against Science itself (Beck, 1992). Thus the truth, not only about the Social World, but to some extent also about the Natural World, has been relativized. The perceived truth about Nature and Society is evaded. Moreover, other aspects of Modernity imply that the public has great opportunities to acknowledge this. Critical Science is no longer a matter for scientists alone. It is a basic resource in the power struggle in Society (Beck, 1992). At the same time the assumed clear difference in esoteric expert knowledge, between professionals and lay people is tendentially obscured, and when the user of the services meet with the professionals the insight is often reversed (Giddens, 1991). The power aspect of the Truth Code[2] (Flyvbjerg, 1991; Foucault, 1979), which is of outstanding importance for professionals, has been undermined. Knowledge, even truth in general terms[3], is being deinstitutionalized, although, in spite of the education boom in the welfare states, the ability to use knowledge has not spread evenly among laypeople. Since I am using an analytical part model this admittedly important question of equality is not discussed further.

The Changing Context and Decentralized Welfare

The above mentioned structurations have changed the context of state welfare service. Moreover, in addition to the legitimation problem of the State and the fiscal crisis, we can, with reference to the notion of disappointment and to the new self-identity, expect overt dissatisfaction with the quality of state services, to the extent that bureaucratic and professional power dominate as steering media.

Governments in many countries have willingly or unwillingly acted upon the new context, primarily using economic measures as constraints, and as steering media. Privatization, and direct user payment, competi-

---

[2] Foucault (1979) explores the power-knowledge nexus:"It is not the activity of the subject of knowledge that produces a corpus of knowledge, useful or resistant to power, but power-knowledge, the process and struggles that traverse it and of which it is made up, that determines the form and possible domain of knowledge" (p. 28).

[3] Including traditional and authoritarian sources of truth.

tion and free choice, have dominated the political agendas. The main impetus is an assumed inefficiency of the public sector. Conforming to a slack point of view (Cyert & March, 1963), the Danish government has utilized small annual percentage reductions of the budget. The measures to provide service for money, have been combined with a certain degree of decentralization (Gyford, 1991; Batley and Stoker, 1991; Burns et.al, 1994).

All in all we envisage a situation with increasing decentralized autonomy, including the street level production of welfare services, corresponding to a greater decoupling of the more detailed guidance and control from the political and administrative centre - and this under tight budgetary constraints.

Applying Giddens´ notion of the duality of structure (Giddens, 1984), the above mentioned autonomy at the mediating level should no longer be seen merely as a constraint. Under the new conditions it must also be appreciated as an enabling quality, which could be activated in a decentralized restructuring of the welfare institutions. The mediating subjects are in a position where they, at least in principle, are able to act reflexively and flexibly, in order to provide users with service for their money. As subjects with power and autonomy, the mediating subjects must legitimize their existence, partly independent of the State.

The legitimation of the qualitative aspects of the Welfare State tends to shift to a decentralized level, and eventually filter down to street level. The shift is going to take place in kind, implying a change away from production for the institutional environment, to a re-emphasis of production for the technical environments (Meyer & Scott, 1983).

In reality, however, we cannot readily assume the existence of a non-complex Modern Condition, in and around the Welfare State. Most of our institutions are dominated by traditional norms and values. The clients are thus not perceived as users with a modern self-identity, by most professions. The professions views are more traditional, and they are still deeply engaged in perfecting their project of professionalization (Hugmann, 1991). Often 'raw material' is a more real description of professionals´ perception of the users.

THE USER-SUBJECT AND THE DIFFERENTIATED MODEL

Nevertheless, decentralization and the establishment of the new power bases do not automatically (re)produce user neglect. Moreover, the organizational structure may be actively altered, and the performance of the could be the decisive party in turning the decentralization to their own advantage. In the case of the emergence of a user subject as an authentic functioning party, it implies a modified organizational structure, a differentiation.

Instead of the traditional direct line of command from the State, via intermediating professionals to clients I would suggest a triangular relationship, connecting the three parties; the political level, the professional, and the users. As the term parties indicates, the users are at this point of modernity present and actively participating in the interpretation of their needs and values. However, I am, not suggesting a self-contained third alternative. The organization and its professional and street level staff should still be held accountable for the service delivered. In the model the users in broad terms representing civil society conduct a supplementary and limited influence on the welfare system.

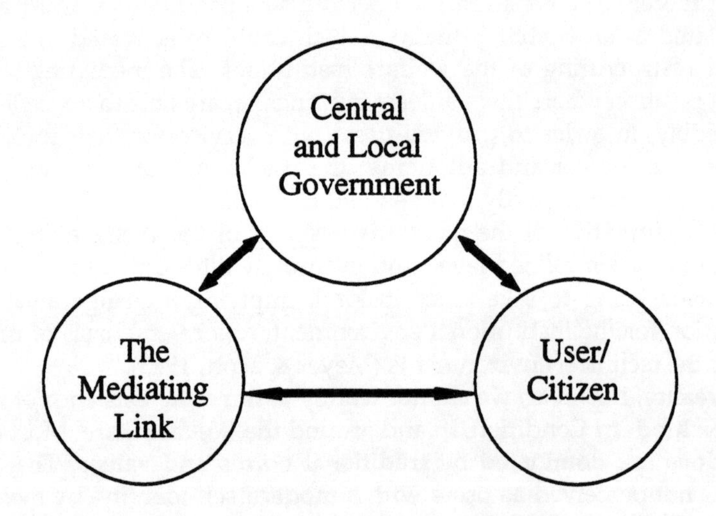

Such local user subjects may be limited by time and in space, but in contrast the body[4] from which they emerge, civil society, is not. Even when the user subjects eventually dissolve as subjects, they may have changed the organization, and perhaps also its greater context. According to Hirschman, their eventual success in changing the structures and the rules could be the very reason for their dissolution. They may "contain the seeds of their own destruction", as Hirschman writes (1982, p.10). Given

[4] The concepts Civil Society and Life-World are used in a non-specific overlapping way in this article.

the differentiation of society, other modes of civil society manifestation are possible; e.g. social movements and civil disobedience (Cohen and Arato, 1992; Habermas, 1987).

THE DANISH AND THE SCANDINAVIAN CONTEXT

The Danes undoubtedly have a passion for equality which has influenced our two great national projects: the farmers' reaction to the agricultural crises at the end of the nineteenth century, where the Danish farming was changed by the Farmers' cooperative movement. Later came the development of the worker movement and the emergence of the Danish Welfare State. (Andersen, 1986; Hofstede, 1980; Borish, 1991).

From the beginning of the 1980's the debate and the search for a new and restructured Welfare State accelerated. In Denmark, the initiative was taken in services related to the elderly. The motive was largely economic and related to fiscal problems, although the substance of the discussions largely concerned the internal quality and the distributional effect of the welfare expenditure. Moreover there was criticism of the old age institution, from the point of view of the micro physics of power (Foucault, 1979, 1982). In a Parliamentary Commission report it was made clear that a change from a financial point of view was urgently needed (Ældrekommissionen, part 1-3, 1980-82). By that time development had already begun at the municipality level.

The search for participatary alternatives in the Danish welfare provision is comparable to the development described in the UK by Hadley and Hatch (1981), of local participative experiments. In the UK, this was been suppressed for more than 10 years. This was not so in Denmark.

PARTICIPATION IN SERVICES FOR THE ELDERLY

Since middle aged workers were first offered the opportunity for an early retirement, and thus to opt out of employment at the age of 60[5], a new social group emerged; the elderly, the *60+ population*. This group shares the common fate of freedom from work until they die, "Die späte Freiheit" (the late freedom) as Rosenmayr calls it. This is a context he most excellently explores in his book of that title. Rosenmayr characterizes the later part of the life cycle as reflexively lived life. Of course the identity of the elderly may be bound to the past, but they certainly have a future - some

[5] The retirement age for both men and women in Denmark is 67.

elderly may be bound to the past, but they certainly have a future - some of them an extended future life span of 25-30 years; the 'condition of late freedom' (Rosenmayr, 1982).

In my view, Rosenmayr shows that the elderly are the first substantial group in society who are flung into the modern condition, a position of having to adopt the new self identity, and become their own project[6].

> "Freiheit hiesse also Verfügbarkeit über Zukunftschancen und durch Planung rückvirkende Erweitung der Handlungsmöglichkeiten der Gegenwart"(p.254)[7].

However, the new 'early'(already from 60) late freedom, which have been formed around the 60+ group, does not automatically cause emancipation of the individual or the group. Yet emancipation does emergs. In conclusion, there are good reasons why the elderly should become predominant in the participative user wave in the Welfare State.

The Danish Center-Right coalition government of the 1980's, started the process in their so called Modernization Programme[8]. Although commencing with a vociferous public declaration of privatization, the programme later turned towards greater user responsiveness and influence, including both exit and voice options (Hirschman, 1970). In a 1993 memorandum from the Ministry of Finance the above described triangular model: State, professionals, users, was actually used.

The tripartite relationship between those parties gives rise to three alliances; the traditional alliance between the political level and the mediating link including the professionals, who share the interest of legitimation; the resource alliance between the professionals and street-level bureaucrats and the users; and the new alliance between the political level and the users, with quality of service and effectiveness as the shared interests.

The relationships also constitute a trilateral check of power. The patterns of alliance can be expected to shift over time, as happened to services for the elderly. It all started with a loose alliance between the political

---

[6] Of course, young people also experience deepening of the modern conditions, although not as dramatic a change. It can be said to be with them from the outset, and not in combination with a final freedom from the labour market. Their material basis for life is not confined in the same way as it is for the 60+-group of senior citizens.

[7] Freedom implies the availability of future life poissibilities, and, with back reference, through planning the expansion of the present scope of action'.

[8] The programme succeeded in its aim of decentralization, whilst maintaining the overall budget constraints.

level and the user, in the 1970's, when the transformation of services for the elderly began. In the 1990's the discrepancy between the resources and the basic needs of the frail elderly has increased. As a direct consequence the elderly took action by reestablishing the resource alliance. They founded a movement; the so called "revolt of the C-team". Although initiated by independent professionals and intellectuals the 'revolt of the C-team' soon developed into a rather large grassroot movement led by the elderly and their relatives. The movement organized demonstrations outside Parliament. They committed civil disobedience when the Council of Copenhagen closed down a number of homes for the old and disabled. They also organized petitions to the Government. The movement's aim was to influence the central decision-making level. The 'revolt of the C-team' operates as a pressure group on the input side of the welfare state. They do not act in accordance with the user role.

On the local level, the same alliance was motivated in my home city of Aalborg. In connection with a major restructuring of the care for the elderly, the elderly protested and organized a local citizen's action group, called "old age with dignity". From the start the group was strongly influenced by the staff in the sector who opposed the organizational programme of the municipality which undermined the power base especially of the nursing staff. In a public meeting, the elderly - with one exception only - deplored the new model in support of the staff. Since the elderly criticised on behalf of the staff and hardly mentioned their own needs, it was easy for the political level not to accept the claims of this group as an authentic expression of the interests of the elderly.

Of course it is essential to discuss the question of taking hostage and coöpting (Selznick, 1949) of the users. In reality are the citizens developing a new decentralized user channel of participative influence, or is the staff getting access to yet another channel of influence? Before theorizing further on power bases and asymmetries of parties, the question will be approached empirically by a case-study.

THE CASE: USER INFLUENCE IN A SUPPORTIVE SETTING

I now turn to a case where the condition of real influence has developed over a period of about four years. The superior rationale of the project was conceived in general terms from the outset, as was the criteria for success of user influence; and a constitution through a hermeneutic trial and error process succeeded. The dynamics were related to an awareness of the unintended consequences of the behaviour of the actors (Giddens, 1984). The success could be said to be coincidental, yet a path was searched for throughout the process.

In the end the creation of a user subject constituted the interest of the users and raised their consciousness as a prerequisite for their emancipation, and the simultaneous disclosure of a state of double false consensus (the concept is described below) between system and user.

I was connected to this project as a consultant. My role was to evaluate aspects of user influence. The field work was funded by the Ministry of Social Affairs, and by independent funds.

ORGANIZING INTEGRATIVE CARE AND USER INFLUENCE IN 'KLOSTERHAVEN'

The object of the case-study was a care and service centre for the aged called 'Klosterhaven' in Viborg, the historical capital of Jutland. In 1982, the Viborg Municipal Council decided to build a new nursing home to facilitate the provision of better institutional care. The result, however, was the opening of the integrated center, Klostehaven, on November 1st, 1985.

Klosterhaven serves an area of 10.000 inhabitants, including 2.000 above the age of 65. The total population in the municipality is approximately 40.000 and stable. Klosterhaven was established as an experiment, and the most important and ambitious goal was that of user-influence. At the same time it was a divisionalization (Mintzberg, 1983) of the care in the town. The "divisionalized" unit "Klosterhaven" was organized under the home help service, and the staff was recruited from or by this by bringing with them a non-institutional care culture.

The following figure indicates the structure of the organization and the line of command between the social services and Klosterhaven.

Klosterhaven offers all the important and common forms of nursing, care and service to the elderly, e.g. nursing home facilities, rehabilitation, day care, preventative health care training, home help, out-reach work, social and cultural activities, meals, occupational therapy, physiotherapy, clubs (e.g. theatre and card clubs), and café facilities with live entertainment. There is room for 21 residents. In 1989 the staff comprised: 96 home helpers, 15,5 district nurses, 4 assistant nurses, and 6 people occupied in preventative health care training, and cultural activities (Wadskjær, 1991).

Viborg Department of Social Services

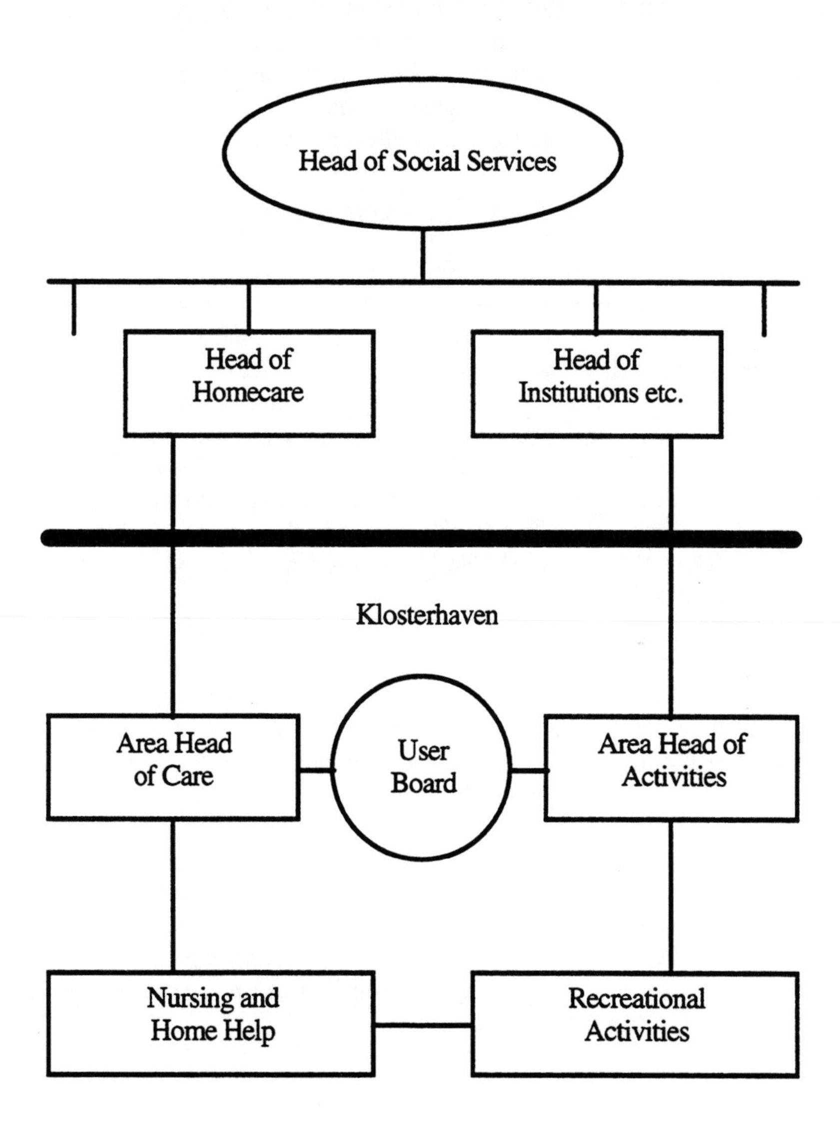

There was a devolved management, with a nurse as leader of the nursing care and home help service, and an occupational therapist as leader of the other activities. The two leaders were the only staff members on the User Board, and only the five user representatives had a vote.

The user representatives are elected at public meetings for old age pensioners usually with 100-200 participants. Two and three user representatives, respectively, are elected in alternate years.

An Unfinished Alternative

No plan for the experiment was worked out before it officially began to function on November 1st, 1985, two weeks before the local elections[9]. Klosterhaven became an event in the election campaign, which caused a premature start with extra chaos. Consequently, Klosterhaven was established as an unfinished experiment, solely on the basis of a number of ideas and philosophical principles on how to treat old people properly, and on how to redistribute power. Mathiesen defines the relationship between the dominating system, or organization, and "the unfinished innovation" as a "competing alternative". Two conditions constitute the unfinished character: 1) the new challenge or message must be a "foreign" element in the system, instead of being woven into the existing organizational system; 2) the content of the challenge, or message, must be sketched or indicated rather than made explicit (Mathiesen, 1971). The crux of the matter is the paralysing effect on the established structures. The unfinished mode of action dispenses with the duality of structures by making a break in the reproduction over time. From the outset, Klosterhaven was to some degree protected from an attack by the established and powerful actors. Giddens´ duality of structure is of little help in the problem of deliberate and free innovation in social relations. Actually, he gives all the good reasons for constraining the future structure. However, as Connel maintains, practice can be turned against structures (Connel,1987). And this is exactly what Matiessen does in his construction of the duality of the unfinished, the decoupling of present structures as an organic aspect of the innovation.

[9] In Denmark local elections; i.e. elections to the 14 County Councils and the 275 Municipal Councils take place every four years.

THE CONSTITUTION OF KLOSTERHAVEN

Klosterhaven as a new organizational unit initiated a constitution process, and because it was an unfinished entity, the process started as a challenge to traditional codes. the point of view of the dominating staff, the nurses, the following problems occurred: the start was premature; the Association of Nurses never took part in the discussion of the new concept of care (which offered, in the minds of the nurses, no clear guidelines for implementation). In their opinion, the number of nursing home beds was reduced too radically; and district nurses were also made responsible for institutional care. Because the development of user influence depended upon the process of organization, this is outlined with emphasis on the political issue in each of the development phases. Having a hermeneutic momentum the constitution took the path as shown below.

No	Phase	Characteristic
1	The start	The start was chaotic due to a simultaneous planning and implementation of activities.
2	Conflict	A conflict with the administration began when staff reacted forcefully against external influences. The experimental unit against the bureaucracy.
3	Adoption	A intermediate strategic phase. Working out a new agreement or understanding between the experimental unit and the Social Welfare Department.
4	Concentration	A phase of concentration, in which the experiment could concentrate on efforts to fulfill the goals.
5	The end?	Institutionalization. Still unfinished?

Phase One. In an unfinished experiment, chaos is to be expected, and this one was a disharmonious chaos. The possibility of *a* growing user interest and user consciousness was less obvious. No one was "present or stood still long enough" to provide a basis to act upon. The users shared the common condition of chaos. The chaos during the first nine months of the first year was reproduced as most of the professionals, the nurses, continued acting in accordance with their former codes of practice. They did not themselves as accountable for the present chaotic situation, so they did not sanction any constructive effort to organize Klosterhaven. They acted in order to make Klosterhaven a non-competitive alternative (Mathiesen, 1971) with the intent of preventing it from being considered a realistic alternative. In many ways time work against them. But eventually they had to realize that they were actively reproducing the chaos. They were demotivated, and their behaviour was a perhaps unintended revolt against the political directives. Their conduct also had severe and unintended repercussions on the staff itself. Staff-members became ill, and some resigned in despair. The phase ended because the nurses reflected on the situation - set off by the loud exit of a colleague´ - triggering the likely outcome: "In the end we will all have to leave".

In the opinion of the users, the staff took too many liberties, but they did not think it was wise to voice their opinion, did feel strong enough to do so.

Phase Two. After nine months, the unit had reached a point of organization which made it possible to react to the outside world. The dominating reaction was to show hostility to the Department of Social Services as the entity responsible for the experiment and for the premature start. The users shared the common themes of conflict, which were questions of resources and autonomy.

The development in Phase Two indicated a well known problem; the fusion of users and professionals against the bureaucracy. In this phase the users developed a consciousness which was shared with the staff, strongly favouring the experiment. It began to look rather like the "traditional process" of an experiment with an alliance being formed between staff and users.

First, the conflict with the bureaucracy deepened. At some point in the conflict phase, the users chose to confront the bureaucracy directly. It was a logical decision when we consider the internal discourse. The users lost this battle, and thereby triggered a sequence of lost battles for the experiment. Within a few months the main actors realized that a continuance of the conflict would have the unintentional result of ending Klosterhaven as an experiment. According to Mathiesen, it would finish the project. Reflecting on this prospect, the leading actors in Klosterhaven and the lower echelons of the Department decided to start a new phase. Because they were taught to think in phases they called it 'the phase of adoption' from the start.

Thus *Phase Three* was a phase which involved dialogue. Our ("our", because I was deeply involved in the process) intentions were to influence the top of the administration and the politicians, and to convince them of the mutual interests in the experience; implying the necessity for ensuring experimental working conditions. Our aim was to change the traditional bureaucratic attitude of the Department of Social Services into a modern reflective attitude - at least in relation to Klosterhaven. For the first time, the Department was deliberately being challenged to reflect openly on the scope of the experiment, and to acknowledge that the conditions for success were in reality depending on the modernising of conditions in the relationship between the Department and Klosterhaven.

The 'adoption' succeeded. The phase culminated in a 'mid-stage' conference, where the bureaucracy, the politicians, the nurses and other relevant persons and groups committed themselves in public not only to the rationale of the experiment, but also to 'fair terms' of implementation. Klosterhaven was relieved from some strict bureaucratic rules and from some detailed budgetary control. In brief, Klosterhaven obtained experimental working conditions. In the vocabulary of the triangular model, the political level established the necessary limitation on the steering medias of power and money. Klosterhaven was partly decoupled from institutionalized rules, and survived as an unfinished experiment on a higher level, so to speak. The question about real influence was now a question of the professional codes against the users media of influence (Parsons, 1967). For the experiment an autonomous scope of action was provided within the bureaucracy. But would this leave room for user influence as well. How would such influence be constituted on the part of the users?

Phase Four. This phase emerged as the result of a long and reflective process and, in the end, a liberating one for all the actors in Klosterhaven, including the users. The user representatives on the Board and the staff were no longer occupied by conflict with external actors, and could now begin to fulfill their mission. User influence and the new concept of care, could succeed relatively independent of one another. After more than two years, the experiment could begin to function on 'fair terms'.

The identity of the users had partly merged with that of the staff. They had cooperated in a successful alliance against the bureaucracy. The big question was now whether or not they would be able to differentiate their more specific user identity (in opposition to that of the system) in this supportive organizational context where the management and at least some of the nurses shared their goal of furthering user influence.

The question now became: could user influence become a reality in a supportive setting with a predominant agreement on the goal of participation? Or were the users facing a coöptation?

In respect to the creation of user identity, the most promising instances were the discussions on the competence of the Board, and user participation in the handling of complaints from other users and their relatives. These are occasions where the differences of interest between the users and the system are often exposed, and in a rather obvious way.

In Phase Four, negotiations took place. Reasonable decisions were made; e.g. the agenda was negotiated and agreed upon; apparently there was no filtering of interests; there was implementation of decisions; and the discussion on the Board was oriented towards consensus but not biased by strong media. It appeared to be a continuing success for user influence. At least I was unable to observe signs of coöptation (Haug and Sussman, 1969). There were almost no differences in the goals.

The Creation of a User Subject and the Emancipation

At that point in time I was hesitant to draw conclusions. What about Lukes´ (Lukes, 1974) third dimension of power, and Parson (1967) and Luhmann´s (1976) media and codes. Or were the users like Bateson´s frog (1972) slowly being "boiled" to become part of the system - a reification of the professional codes? Were they still an authentic manifestation of civil society? What kind of dualities of structures were constitutive if any? (Giddens, 1984).

In an article on Klosterhaven (Wadskjær, 1988), I suggested the creation of a pure user forum. Of course the staff as well as the users read and discussed all my articles, including that one.

In the forth year, the user representatives on the Board actually formed such a forum. They realized that the Board had become overcrowded in their attempt to broaden participation beyond the five elected members. According to the judgement of the users, the Board could not function under those conditions. Instead they created a forum called 'the Agenda Meeting' - with open participation, but only for users - for discussions of user issues. They institutionalized, on a fragile bottom level, a user subject which was limited in space, and also perhaps in time. The creation of such a subject was not the ambition of the decision, instead they attempted to create an alternative way of broadening the direct democracy, because the first method had unintended drawbacks on its capacity to organize their influence.

The ultimate emancipation followed a few weeks later. In one of the first agenda meetings the users discussed their priorities for the forthcoming Board Meeting. On this occasion, a question of budget changes was of special importance to the users, so they concluded that the issue should on the agenda for the next meeting. The competence of the Board was quite explicit on this matter. As usual one of the user representatives on the Board, who happened to be unexperienced, met with the leaders of Klosterhaven, to decide on the agenda for the Board Meeting.

The issue was missing on the official agenda. The leaders had suggested to include the issue only as a point of information. Consequently, the issue could not be decided upon. In fact the leaders had already decided the issue in what they honestly believed was in the best interests of the users.

At that time I had not been present at the Board Meeting for a year. When the more experienced users saw the agenda they became very upset and invited me to participate in the meeting.

This was a revolutionary event: the disclosure of the difference between system and user. Briefly stated, the users informed the leaders that they were wrong, in substance as well as in procedure. Moreover, the users considered leaving (Hirschman, 1970) the whole project, but obviously not silently (Barry, 1979)[10]. The threat was not explicitly commented upon, but it was clearly felt not to be a flattering event to be reported in a research report.

On the surface, it looked as if a deliberate manipulation on the part of the leaders was being practised. In fact it was not. There was no deliberate concealment, simply because in the minds of the leaders there was nothing to hide. The process has to be interpretated as an expression of 'normal practice'! 'Normal' must have implied a relation of double false consensus between the system and the users. Both parties saw themselves in opposition to the municipal bureaucracy. This case exhibited that the two parties had developed and institutionalized a consensual mode of negotiation. Previously both parties had worked under the tacit perception that they had common interests only in the negotiations of the Board. This was how the leaders saw their own conduct. Consequently, they were shocked and their feelings and professional integrity hurt. In conclusion, the user representatives and the two leaders had created the consensual mode themselves. Both parties believed that they were in opposition to the bureaucracy and fell victim to their common creation. That is, until the advent of the user subject.

Before any further discussion on the emancipation of the user subject and the condition of double false consensus, I discuss the positive consequences of the event. The two female leaders - perhaps due to the very fact that they are female - were able to learn and adapt quickly to the new conditions. The user representatives declared that the Board, in their opinion, never worked better since this event.

Conditions for user influence apparently had been established and were working. There were two main parties with separate identities. One was connected to Klosterhaven as an entity, the other was anchored in the user subject. Both were acting reflexively in their mutual relationship. The

[10] As Barry rightly pays attention to, exit and voice are not real alternatives. The alternatives are instead exit/stay and voice/silence. Exit and stay combine with both voice and silence.

professional codes and the system powers were visible, and the parties could and did deliberately create room for influence. The implication of the disclosure of false belief was a transformation of actions in the proper context, a relationship which Giddens maintains to be non-contingent, as is illustrated above. (Giddens 1984, p.340)

Discussion

By implementing the Klosterhaven project the political system had already decided to change a minor segment of the Welfare State to the strong advantage of the users. Thus it could be said that, 'if there was any place on earth without a need for increased user influence it must have been Klosterhaven in 1985'. A possible presumption for the emergence of the user role is the municipality's elaborate provision for the elderly. By thus making superfluous the elderly's citizen role, a scope of action for the creation of a 'different' role, the user role, was created.

Returning to the case of Klosterhaven, where the users over time developed their competance and roles. There was reservation as to the content of the role, and the constitution of user identity and user interests until the users detached themselves from the context by establishing a pure forum (the Aagenda Meeting) for user deliberation. From that forum the critical turning point; the constitution of the user subject. In this case the user subject turned out to be, I will maintain, the necessary condition for the constitution of user interests and for user influence, conditioned as it were on provision of adequate space. This brought about by a supportive setting and the neutralization of outside pressure during the adoption phase. 'But is it easy to evaluate the possibility of a user influence model under such convenient conditions?' On the contrary; it is the only possible condition in which the model can be evaluated.

Let us contemplate the negative scenario. What implications would follow if user influence did not function in a supportive setting? A differentiation of the Welfare State by the introduction of user influence would then most likely not work. In the Welfare State active users would then either be coöpted or dominated by institutionalized power. If other direct and stable manifestations of civil society do not function, the Welfare State is likely to restlessly oscillate between state and market dominance[11]. The importance of adding the medium of user influence to the power and

[11] Hirschman reaches that very conclusion in his analysis of disappointment; the shifting involvement in private interests and collective action. However his analysis is grounded on the Public Choice paradigm (Hirschman, 1979).

money media of guidance and control, is the realization for the possibility of a third aspect of involvement. This can interfere with the oscillation between market and State. This third aspect implies - at least in a Modern context - a reflexive model and the involvement of Civil Society; here taking the form of reflexive user subjects.

Is it possible to generalize? Klosterhaven is only a single case, however it is neither difficult nor complicated to form a pure user forum; the embryonic state of the user subject. The above mentioned inclination to shift away from collective public involvement could cause problems, not for user influence as a possible medium, but for the possibility of the emergence of a concrete differentiated welfare model. However, this apparently is not the case among the elderly in Denmark. The Welfare State is an inseparable part of their own emancipation. To the elderly, at least in Klosterhaven, the Welfare State Project is not yet finished, and the opportunity to engage in its perfection was welcomed.

The problem of coöptation, conceived as a process, is located in the medium of power and in the professional codes of truth, and not least, in the code of love (Luhmann, 1976)[12] with affinity to the calling. Because influence is a relatively weak medium (Parson, 1967), other media and codes tend to dominate the dialogue between professionals and lay people. The professional codes function most effectively combined with trust (Luhman, 1976; Hasenfelt 1978), i.e. in a supportive setting. Therefore, the user subject can only be expected to exist or to be constituted outside the system, i.e. whilst the user interests are initially taking form. This reminds us of Marx´ early writings on the proletariat and the Jewish question (Sherover-Marcuse, 1986). From such a position, emancipation may, eventually, proceed.

From the perspective of the system, the necessary space for influence in a modern context may be reflexively established. In the constitution of Klosterhaven, limitations were established on the medium of money, the medium of institutionalized power, and the professional codes on several occasions. In the last case, this was brought about by professional self-limitation, and, in the first two by means of political will (in the case of the adoption phase).

In a time of professionalization of everyone the question could be asked if user subjects are genuine (not influenced by professionals). However, another problem, that of the user segmentation aspect of the model, could be of some assistance. Other professionals are users too, as such

---

[12] Luhmann suggests that we consider love and knowledge as media, which facilitates communication.

they have user interests. Professional groups are often in conflict and are reluctant to recognize each other. The authenticity of the user subjects are thus sustained by the user segmentation aspect of the model.

Recapitulating the extreme qualities of the case and the explanatory impact of the theory of media and the user subject, with reference to Civil Society and Life-World, I maintain that the general conditions for user influence have been established.

System Constraints for User-Influence: The Schools

In this section I briefly discuss the Danish public sector schools. The intent is to disentangle some of the conditions for user influence on the side of the system, including the impact of the dominant profession. The medium of influence is unsatisfactorily functioning from a user point of view. In the public sector schools a model with user representatives is established by statute but there are barriers in relation to the constitution of user subjects.

In Denmark civil society is both traditionally and formally accountable for the education of the young. The parents are permitted to teach their own children, but that is unusual. Instead, we have a substantial sector of private (we call them 'free') schools (approx. 11%). In both respects Denmark is atypical in comparison with the rest of Scandinavia. The Danish state is compelled by law to provide 85% of private schools funding.

The public sector schools have a 20-year tradition of user influence, i.e. the influence of parents. In 1990, in the general spirit of the Conservative-liberal Modernization Programme, the school councils (i.e. parent-teacher fora) became governing bodies. In the change the Boards have maintained a clear majority of parents. They thus are comprised of 5-7 parents and two staff-members, with the principal as executive secretary. Moreover, the municipalities may decide to allow two pupils to be elected on equal terms with the other members to each of the Boards.

After the first four years the new Boards appear to be a failure in light of the ideal of participation. Participation in the 1994 School Board election was only about 75% of the 1990 level.

On the question of user influence the teachers make up a traditional semiprofession. The Danish teachers have generated one of the most impressive and successful semiprofessionalizations. The degree of closure (Weber, 1978) and systemness is especially remarkable

Consequently, the teachers have developed the mind of a conqueror. They are definitely not supportive of sharing of any power with the parents (Windinge & Pedersen, 1991). In that respect, the public sector schools are traditional, on the verge of fundamentalism. Usually there are strong cultures in the schools, and there are also excellent opportunities to reproduce them.

In the schools the difference between user identity and that of the teacher is unmistakably clear. Nevertheless it does not render the establishment of a user object irrelevant but the function differs from the previous case. Collective action in the schools is more a question of a power struggle and of self-protection. Parents feel that their children are dependent on the teachers, who may take revenge on them. In my judgement, such revenge is not at all common in Danish schools, although fear of revenge is not totally unwarranted.

Those who against all odds try to influence their children's school, should aim at modernization by challenging tradition in direct discussions with members of the faculty. The very questioning and subsequent reflexion and defence of a tradition erode its very existence as such.

The parent majority on the Board should use their statutory powers and make decisions. A few such cases have come to my knowledge. Typically, the cases result in a "civil war" between the parents on the Board and the teachers. The latter are often in alliance with the principal and with the local branch of the Teachers´ Association. In other cases, there has been filthering (Windinge & Pedersen, 1991) and even non-implementation. The principal functioned as a buffer by simply refraining from implementing the decision or ignoring it when colleagues neglected the decisions of the Board which had no management functions and thus can not oversee the day-to-day practice. In my home city, Aalborg, a principal was discharged on this account. For three years he refused to abide by the statutory budgetary competence of his School Board. He apparently did not approve of the law. He also underestimated the ability, and the motivation, of the user subject to act.

Very little independent qualitative research is published on the issue of participation in the Danish schools, though a few are under way (Cederstrøm,1991; Sørensen, 1994; Rasmussen, 1992). Even studies which have been undertaken by the parties whose legitimation is at stake, show great problems (Cranil, 1994). The emerging picture appears to be one of formal coöptation, meaning symbolic legitimation without any real influence, based on the organizational power of closure, with the professional truth code playing a minor role, and with the love code barely existing.

The existence of real alternatives for exit (Hirschman, 1970) provide the parents and pupils with the opportunities to combine voice and exit. This is a safeguard in cases of revenge. According to Hirschman exit may have a repressive effect on voice. No one is compelled to have their children taught in any particular school, or any school at all. Ready at hand is the argument that 'if you don't like the smell....'.

In some of the private schools, the practice appears to be much more favourable regarding user participation (Sørensen, 1994).

In the public sector schools, the battle continues. Survey and field studies appear to reveal increasing Board involvements of the parents who are also teachers, in addition to those parents who are married to a teacher (Cranil, 1994). Their presence on the Boards in many cases constitutes a teacher biased majority. Even when they do not have a majority, they will very likely, however unintentionally, unreflectedly apply the professional codes of truth and even love, because their status as parents, on equal footing with the other parents, will make room for their influence. Thus it demolishes any attempt to form pure user fora and user subjects.

Male spouses - often other professionals - seem to dominate the local discourse, and one may ask about the profession of the head of the nationwide parent organization: "School and Society". Indeed, she is a teacher by profession. It is, in fact, difficult to find any pure parent forum for discussion and creation of a user subject. Logically the teachers are suspicious when the parent Board members meet or talk outside the Board Meetings. Some principals even declare that such practice is in contradiction to the statutes, thus using the law as a medium. So, there are great difficulties in constituting local parent interests, both at the school level, municipal level and nationally.

In conclusion, in the public sector schools there is little room for influence, even if it were possible to meet the condition of forming user subjects. The systemness, the homogeneity of the task, the 'hostage structure', the inclination towards union power, and the system's use of the exit option, all comprise conditions which are unfavourable to user influence.

The organizational field differs to a great extent between care for the elderly, and school education. Only the latter has been through a successful process of structuration followed by a successful closure of occupation (Weber, 1978). However, teachers still comprise, perhaps permanently, an aspiring profession. A process of professionalization which according to New Institutionalism, is liable to entail tension. In a case of creating an organizational field DiMaggio discovered a collective behaviour which had a great resemblance to that of the teachers;

> "What is striking, however, is how little conflict occurred **inside** organizations and how much was played out **at the level of the field**". (DiMaggio, 1991, p.268).

The success in structuring and forming a closure, or the failure of these, as in the case of the nurses in the field for the care of the elderly, appears to be but one single variable, which correlates with the creation of space for user influence.

Concluding Remarks

In the current restructuring of the Welfare State, the case of Klosterhaven supports the possibility of a differentiation, in accordance with the general trend of decentralisation. The necessary condition for constituting user interest has been shown to be the creation of a user subject - which is limited in space, and perhaps in time. When speaking of user subjects, I am not talking about a new movement, but I do see the subjects as manifestations of civil society.

In the case discussed, there must have been general conditions present in part of the system, in order to secure success. These conditions are for the time being missing in other sectors, e.g. the school sector. Broadly speaking, space for influence is missing. However, I have not explicitly analyzed the matter of structuration of different fields or the professionalization processes. The analysis has been focused on the users, resulting in the coining of the concept of user subject whose practical implication is the constitution of reflexive collective acting user groups, e.g. in Boards and Councils.

When the space for influence is established, the functioning of the triangular model involving civil society in the form of a user subject is confirmed. In the establishment of space the will of the political system to limit and to balance money, bureaucratic and professional powers which oppose influence, was shown to be one important factor. Another was the self-limitation on the part of, admittedly from a Weberian point of view, the weakly organized professional nurses.

The education sector is an organizational field with a high degree of institutionalization, and with strong closure. In that sector, there are problems even with the constitution of user subjects. Furthermore there is no space for influence. Henceforth the modern active users are engaged in a countervailing power struggle, or rather a struggle for the unveiling of the power-knowledge nexus. Perhaps they are hoping for political mediation, or self-limiting behaviour, from the associations of the professionals.

References

Andersen, B.R. (1986). "Rationality and irrationality of the Nordic Welfare State". In S.R.Graubard(ed.), *Norden: The passion for Equality*. Oslo.

Barry, B. (1979) "Review article:'Exit, Voice, and Loyalty'. *B.J.Pol.*S. 4, 79-1.

Bateson, G. (1972). *Steps to an Ecology of Mind*. Ballantine, N.Y.

Batley, R and Stoker, G. (eds.)(1991). *Local Government in Europe-Trends and Developments*. MacMillan, London.

Beck, U. (1992). *Risk Society-Towards a New Modernity*. Sage, London.

Borish, S.M. (1991). *The Land of the Living; The Danish folk high school and Denmark's non-violent path to Modernization*. Blue Dolphin. Nevada City, California.

Burns, D. and Hambleton, R. and Hogget, P. (1994). *The Politics of decentralisation*. MACMILLAN. London.

Cederstrøm, J. (1991). *Samtalen i skolen - en undersøgelse af forestillinger om og forventninger til lærerens kommunikative rolle.*Unge Pædagoger. København.

Cohen, J.L. and Arato, A. (1992). *Civil Society and Political Theory*. The MIT Press, Mass.

Connel, R.W. (1987). "Gender & Power." *Polity*, Cambridge, UK.

Cranil,M. (1994). *Decentralisering og selvforvaltning i folkeskolen*. Undervisningsministeriet, København.

Cyert, R.M. and March, J.G. (1963). *A Behaviour Theory of the Firm*. Engelwood Cliffs, N.J. Prentice-Hall.

DiMaggio, P.J. "Constructing an Organizational Field as a Professional Project: U.S. Art Museums, 1920-1940." In:Powell,W,W and DiMaggio, P.J. eds.(1991) *The New Institutionalism in Organizational Analysis*. The University of Chicago Press. Chicago.

Flyvbjerg, B. (1991). *Rationalitet og Magt*. Akademisk Forlag, København. (Forthcoming in the US).

Foulcault, M. (1979). *Discipline and Punish*. Vintage, New York.

Foulcault, M. (1982). "The subject of Power". In Dreyfus, H. & Rabinov, P. *Michel Foucault: beyond Structuralism and Hermeneutics*. Haverster Press, Brighton.

Giddens, A. (1984),*The Constitution of society*. Polity, Cambridge Uk.

Giddens, A. (1991),"Modernity and Self-Identity". *Polity*, Cambridge,UK.

Glaser, B.G. and Strauss, A.L. (1967). *The Discovery of Grounded Theory - Strategies for Qualitative Research.* Aldine De Gruyter, New York.

Gyford, J. (1991). *Citizens, Consumers and Councils.* MacMillan, London.

Habermas, J. (1990). "The Crisis of the Welfare state and the Exhaustion of Utopian Energies". In: Seidman, S. *J.Habermas on Society and Politics.* Boston.

Habermas, J. (1987). *The Theory of Communicative Action,* volume two. Polity Press.

Hadley, R. and Hatch, S. (1981). *Social Welfare and the Failure of the State. Centralised Social Services and Participatory Alternatives.* George Allan and Unwin, London.

Hasenfeld, Y. (1978). "Client-Organization Relation a System Perspective" In Sarri & Hasenfeld, (eds), *The management of Human Services.* Colombia University Press. New York.

Haug, M.R. and Sussman, M.B. (1969). "Professional Autonomy and the Revolt of the Client". *Social Problems.* Fall 1969, 17(2):153-61.

Hirschman, A.O. (1970). *Exit, Voice and Loyalty.* Harvard University Press.

Hirschman, A.O. (1982). *Shifting Involvements.* Martin Robertson - Oxford.

Hofstede, G. (1980). *Cultures Consequences.* Sage. Beverly Hills, California.

Hogget, P. (1990). *Modernisation, Political Strategy and the Welfare State, An organisational Perspective.* Studies in Decentralisation and Quasi-Markets no. 2. University of Bristol.

Hugman, R. (1991). *Power in Caring Professions.* MacMillan, London.

Le Grand, J and Bartlett,W. (1993). *Quasi-Markets and Social Policy.* MacMillan, London.

Kean, J. (ed.). *Civil Society and the State.* Verso. London.

Luhmann, N. (1976)"Generalized Media and the Problem of Contingency". In Loubster, J.J. et al. *Explorations in general Theory in social Science* 1-2, Essay in Honor of Talcott Parsons, New York.

Lukes, S. (1974). *Power - a Radical View.* MacMillan, London.

Mathiesen, T. (1970). *Det uferdige.* PAX, Oslo.

Meyer, J.W. & Scott, W.R. (1983). *Organizational Environments: Ritual and Rationality.* Beverly Hills, CA: Sage.

Mintzberg, H. (1983). *Structure in Fives: Designing Effective Organizations.* Printice-Hall International Editions. New Jersey.

Rasmussen, J.G. (1992). *Skoleledelse - udvikling eller afvikling.* Kroghs forlag, Vejle.

Parsons, T. (1967). *Sociological Theory and Modern society.* NY.

Rosenmayr, L. (1982). "Die Späte Freiheit", Das Alter - ein Stück bewust gelebten Lebens. Severin und Siedler.

Selznick, P. (1949). *TVA and the Grass Roots - A study of Politics and Organization.* University of California Press. LA.

Sherover-Marcuse, E. (1986). *Emancipation & Consciousness.* Blackwell. Oxford.

Sørensen, E. (1994). *Nye demokratiformer i den offentlige sektor.* Unpublished working paper. University of Copenhagen.

Wadskjær, H. (1988). "Brugerstyring - et eksempel fra ældresektoren" I: Jensen,O.B.(1988) "Helhedsveje". Alfuff, Aalborg.

Wadskjær, H. (1991). *Resultater og brændpunkter.* Alfuff, Aalborg University, Aalborg.

Weber, M. (1978). *Economy and Society.* Volume One. University of california Press.

Williamson, O.E. (1975). *Markets and Hierarchies.* The Free Press, N.Y.

Windinge, H. & Pedersen, E. (1991). "Organisationsudvikling af folkeskolen". *Unge pædagoger* no 8 Dec. København.

Chapter 8

THE FRAGMENTED LOCALITY

Peter Bogason

1. CONSEQUENCES OF CHANGE

The chapters of this book illustrate a number of facets of the changes in local politics. They do so from various vantage points: those of observers coming from three different countries and those of different research approaches within political science and public administration. This concluding chapter intends to bring together an overall discussion of the evidence and the prospects for future analysis.

The idea, then, is to generalize. This is done in full recognition of the fact that many details will be ignored, so there certainly is a strong element of speculation in the following pages. But generalizations and informed speculations serve the purpose of advancing academic discourse, and discourse is the precondition of theoretical improvement.

First, we draw some conclusions regarding the changes in local government and governance. Then we relate the changes to these changes in the Postmodern society, and finally, we shall discuss some possible consequences of a more fragmented public sector.

2. Changes in Local Governance

Where is government in the locality heading? Surely the Scandinavian consolidated model of local government is on retreat; powers are being moved out from town hall to other forms of local representation, and to a certain degree market forces are being strengthened. The reactions of Scandinavian observers vary, but so do the actual developments in those three countries.

The size of local government varies among countries and even within countries. There seems to be no perfect size, neither in terms of participation nor economies of scale, or rather, no country has moved to follow the ideas of either democratic theories or theories of the firm to establish standard sizes of local government, be it rather small "democratic" communes or large "efficient" municipalities. In actual fact, local governments vary from below 1000 to well over 100.000 inhabitants. The general national political stance has been to increase the size of the formal general purpose entity, but recently there has been a positive mood towards encouraging new local ways of participation - undoubtedly creating some activity in the existing organizations to preserve the values that are held high there. They include, of course, a wish to keep the kind of political control that has been established over the years.

From Sharpe's chapter one can infer that if popular participation is desirable, some of the existing local governments are not really able to respond without organizational changes. Such changes might be linked to the accessibility of services. Services provided by the municipality might not be linked solely to the municipality council and its town hall administration, but could be closer linked to users by several means:

- user boards of directors
- districts and neighborhoods
- contracting out to private or non-profit organizations

The philosophies behind these measures differ significantly, but the rhetoric about users and/or customers getting closer is about the same. As a consequence, the countries in general are giving up the old integrated general purpose unit of local government. All the elements above are found in the countries discussed in this book. But the ways they have been introduced certainly differ.

Some analysts (see e.g. some of the contributions in Stewart & Stoker 1989) see the strategies of central government towards local government as attempts towards profound change from relying on a local center for collective consumption (Town Hall) to relying on local economic growth based on indigenous business activities, supplemented by various local public bodies to care for the destitute. The senses of democratic community and social common responsibility that may have been there during an era of a relatively important local government, at least in spending terms, will be gone; the consumers' choices will determine local services. But the observers do note that things may not just end that way quite so easily as the reform discourse indicates; one does not transform governance in the locality overnight. They also point to some democratic values in local boards.

The idea of relying more on business for local growth rather than using the public purse for transfers does not stand isolated within central policies towards local governments. In general, there has internationally been a shift from "Keynesian" transfers towards "Schumpeterian" entrepreneurship within development policies, notably industrial policy, regional policy and technological policy (Beckman et al. 1991). The concept of state responsibility has slowly changed from one of securing welfare by transfers to one of enhancing the prospects of local welfare by reducing the problems that local entrepreneurs meet in their endeavors to start economic growth activities.

This is not to say that social transfers totally disappear. But it means that the interests of public policy-makers are committed to what they perceive as activities of innovation rather than traditional welfare policies, which consequently are not developed or going through a process of sophistication. Only aspects related to job qualification enhancements are being developed to help business entrepreneurs get access to a well-trained work force.

But a number of the traditional welfare services do come closer to the users. Thus

> ... decentralized delivery systems .... offer a way of integrating and better managing the troublesome classes left in the residual welfare state .... It is accepted that these groups need to be given representation within the system, but their position within the hierarchy is clear. ... The fragmentation of the local welfare state helps to confirm these shifts, moving decision making into increasingly closed arenas ... (Cochrane 1991:296).

Cochrane sees this as part of the general strategy towards local corporatism where functional representation of different groups takes place. Elected local governments are one among several channels for local influence, and perhaps has a new role as mediator between interests (Cochrane 1991:299).

For sure, this would be a fragmentation of the local governance system; but there is hardly any evidence yet to confirm a development that has gone so far. And one might wonder why a Conservative government would want to establish a corporate system; a solution mostly preferred by Social Democratic forces within politics.

Anyway, there is a tendency among most of the British observers (for an exception, see Stoker 1991) to see the changes as British phenomena and link them strongly to the Thatcher era (for a round-up, see Cloke (ed) 1992). And indeed they are at face value. But there are comparable trends in other countries, notably in the Scandinavian countries.

In the Scandinavian countries, as we have seen, the local public sector is getting fragmented by neighborhood councils, user boards and to some degree by more contracting out. Signs of local corporatism are not strong. Nor can one say that the fragmented system is a result solely of liberal or liberal-conservative endeavors. There seems to have been consensus in the parliaments about most initiatives requiring parliamentary action[1]. There is general agreement that some of the local government welfare bureaucracies may have had a tendency to grow into less accountable roles and at the same time, clients of services have become precisely that: clients with little influence on the production of those services.

But the approach has been markedly different from e.g. the British who carried through reforms from the top without much interference from the very local governments that were to be changed. The Scandinavians have tried out a host of ways by the free commune experiments. Thus the locals have themselves had a say in defining success and failure. Though some may disagree on the contents of the final decisions, no one can say that a process of command and control has been pursued from the top. The tradition for consultation between ruler and ruled has been kept, albeit under the precondition that ultimately the central government has a strong say (Rose 1990:230-33).

The continuing process of consultation between the central and local government representatives on the basis of local experiments has meant

---

[1] One noteworthy exception is the abolishment of the Danish Metropolitan Council which spans the Capital and three surrounding counties; and was opposed by the Social Democrats and the left wing. Its role, however, was far from the roles of the British Urban Counties.

that the Scandinavian reforms have had a form that did not go against most local government interests. All important concepts - efficiency, user influence, accountability etc. - have been defined more or less in agreement in Scandinavia. Moreover, there has nearly been an absence of clear political party interests in disciplining localities dominated by the non-ruling political parties. Such tendencies have not been of a scale comparable to the Thatcher government.

There is evidence from the chapters in this book that in terms of citizen contact, local boards more than any other kind find it desirable. Montin & Persson indicate that some problems relate to the desires of political parties to maintain political correspondence between central and local political bodies in the commune. Aarsæther and associates have found that people locally are not interested in party political issues; they want local problems to be discussed in the locality rather than as ideological stances thwarting what they perceive as clear evidence.

Bäck has shown that the introduction of neighborhood councils improves contacts between citizens and politicians in the localities, while areas with contracting out and other quasi-market solutions reduce the number of politicians and hence give less possibilities for such contacts.

In a way, this is perfectly logical. Market organization calls for solutions of problems related to services by exit rather than by voice; political organization can hardly survive without voice. It then remains to be seen whether the Swedish system of contracting out really will permit market forces to work. Will dissatisfied customers have the opportunity to go to other service producers and thus use the powers of exit to make inferior producers go bust? This question can only be answered by keeping up the spirit of experimentation and by experimenting with different versions in the localities. If more contract systems turn out to be both efficient and responsive, they should get good evaluations. If organizations governed by users do get efficient and responsive, and in addition serve some democratic channels, they should turn out to be even more successful.

Wadskjær has shown that institutional change is impossible to implement over night. Change takes time; some of the participants have vested interests which are not easily abandoned; some simply are slow in responding to new forms of governance. A complex procedure involving many actors and an open dialogue, furthered by external consultants and/or politicians may be necessary. Professional street-level bureaucrats in particular appear to have the capacity to block progress in user governance.

The positive experiences of past experiments in Scandinavia indicate that new experiments can be carried out and that there is a local understanding of how to deal with the intricacies of terminating experiments that do not reach the desired goals. At the same time, however, it should be understood that success or failure cannot often be assessed at an early stage. Successful change requires hard work and the ability of all parties to adapt. In general, Scandinavians seem able to reach such compromises - the political culture is one of a practical spirit aimed at supporting the public interest, however construed.

The development towards fragmentation is not unique to the countries we have analyzed. One aspect that has not been dealt with is the general tendency to having activities organized by the "third sector" or similar constructs on the border between formally public and private organizations (Hood & Schuppert (eds) 1987). A related observation comes from the USA where during the Reagan revolution of the 1980's diffusion of responsibility was seen by having more to indirect government, third-party government, and government by proxy (White 1989). Contrary to what may have been a popular belief, this development has not meant that topics were in large slices cut from the public agenda. And even if they were, they were hardly removed from public attention. Many organizations which were formally private are heavily dependent on public money (Smith & Lipsky 1993). So the changes have meant that new devices of interaction and control had to be developed to keep politicians and public officials informed of what was going on in a fragmented, but nonetheless in a broad sense public interorganizational system.

3. Postmodernity and the Fragmented Public Sector

The chapters in this book have dealt with various aspects of changes in governance at the local level: Special boards; neighborhood councils and other new channels of local influence. These changes fit to some degree in a more general tendency in the societies.

The changes we are witnessing in the Western societies are often characterized as changes towards Postmodernity. The Postmodern society has many characteristics, and observers do not quite agree as to how many and which are the most important ones. In the literature, titles like *Disorganized Capitalism* (Offe 1986), and *Culture Shift in Advanced Industrial Society* (Inglehart 1990) in a loose way indicate some debate topics although it should be stressed that there is no general agreement about many facets of the development (See Giddens 1990). There is general agreement, however, that the development of the society cannot possibly be explained only from a local perspective. International competition and cooperation determine many of the parameters for local life. Still, most changes are observable in the locality where we live and work.

What is happening? The general tendency in Postmodern cogitation seems to be that the society is becoming differentiated into specialized compartments or units. This development can be seen as a continuation of the trends of the modernity with its division of labor along functional lines. But in the Postmodern society the unifying or centralizing, functional aspects of the process are losing their power. There is no unifying structural principle (Crook, Pakulski and Waters 1992:33) controlling the direction and intensity of the development.

In the production, the assembly line and similar "Fordist" constructs are increasingly being replaced by "Post-fordist" ways of production. Job mobility has grown: first, because higher levels of education have made possible career patterns that involve frequent change of employment; second, because the car and the improvement of the infrastructure have made a job change to another area possible without relocating the family. Largely, the production of standard merchandise is not only for a local market; transport technology and information dissemination have opened up a more global market for most goods. This also means that the consumer hardly feels any special obligation to local retailers when buying standard goods.

Consequently, many of us live in a situation where personal freedom and job mobility is high. We are used to being able to react to dissatisfactory services or job situations by the "exit" option - get another job, buy another brand - although increasingly people also speak up against what they see as intolerable work conditions or poor production quality. Thus they also use a "voice" channel informing bosses and producers that there are limits to what they will accept. At the same time, it should be clear that for certain the modern society still exists; dominating the lives of some of our fellow citizens with much monotonous activity and little individual challenge, once the basic features of the job have been grasped. And with little room for influence on work conditions.

Along the lines of Postmodernity, one could say that in the public sphere, the centralized state is now retreating in favor of more power to decentral political structures or even private (non-centralized, non- corporatist) forms of governance. This certainly concerns us as voters and citizens of the welfare state. When it comes to the neighborhood where we live, the situation is different from the Postmodern production (work) life. Most of us care for the locality where we live, and we tend to be meticulous in our choices of a number of services: day care for our children, school, medical care, care for the elderly etc.. If we do not have a choice due to monopolized supply. We want to be able to "voice" our opinions in order to affect the quality of the service at hand. The exit possibility is not the first one that comes to mind.

So, when it comes to such services, we often find ourselves bound by locality or by the administrative district. In many countries, however, there is a trend towards new patterns of influence - as a reaction to the centralized welfare systems set up after World War II. This book reports on such examples; and there clearly is quite a diversity, but still some general lines can be found. The new pattern, however does not stand alone - as in the production life, there still are many elements of the modern system left: rigid bureaucracies requiring standard procedures and standard information, and little scope for individual adjustment. This is particularly true for welfare cash payments, while services are the ones permitting more choice.

Albæk, is one of the few political scientists who has discussed Postmodernity and the local public sector. He questions the direction of the development towards a post-modern condition since he sees

> no social groups that would rally round the project and carry it forward. (Albæk 1995)

But it seems to me that Postmodernity is not carried or rallied by specific groups. This is precisely the special feature constituting much of the post-modernity; it comes through from many, non-coordinated actions rather than from concerted action. But conscious organizational changes in the public sector may be conducive to a Postmodern development.

I see two lines of theoretical development as particularly interesting for political scientists aspiring to study Postmodern trends in the society - always keep in mind that we are speaking of a tendency, not a full-fledged and well established new order. Old (i.e. modern) times do not disappear overnight.

The first one concerns movements among citizens towards better control of questions which are of particular interest to them. Some of that research is aimed at the role of popular movements, often as reactions to what is perceived as the negligence of public authorities, or even as angry reactions to decisions by public bodies that override some particular interests held by a large group of people. Another line of the research in movements goes into the (re)development of civil society understood as social organization based not on government, but on local networks, norms and trust enabling local cooperation.

The second major direction of research concerns the development of the state and particularly local governments (as creatures of the state) into more differentiated forms of governance. This development involves the creation of several new forms of organization. Some are market oriented as production units bidding for contracts from local governments regarding (typically) waste collection and treatment, fire protection, road management etc. related to infrastructure. Other organizations extend (or restrict, depending on the perspective of the viewer) local democracy to specific services, giving users new voices in determining the quality of services in schools, day care institutions and other types of human services.

The chapters of this book mainly deal with the latter direction of research. But Henrik Bang's contribution indicates that one should keep the first direction in mind, since it goes more into the personal motivations for citizens for collective action, of which local public bodies is but one form. And we should not only keep it in mind, but actively work towards an understanding of action in the locality that transcends the conceptual prison we may find ourselves in when we subscribe to the disciplines relating themselves to the private market, the public state and the civil society.

4. CHALLENGES OF FRAGMENTATION

One challenge of the fragmentation is how to cope with many different types of public service organizations in the locality. We can identify an administrative problem and a democratic one.

For the political leadership of the consolidated communes created up to 1970, the main interorganizational challenge was to deal with mandates of the national government and to get the most out of national-local financial transfers. Most of the negotiation processes related to such concerns were taken care of by the associations of local governments. The main problem at the local level was the interpretation of the mandates. Many local governments handled this as would any bureaucrat: in cases of doubt, ask your superior - in this case, the national ministry. Affairs of the commune were handled in a rather bureaucratic manner - service organizations were handled as subordinate agencies.

For the citizens there were only one or two political bodies like counties and communes to worry about or engage in. The choices were presented at election time, and if one wanted to be more involved, one could use the local press or get involved in party politics. The relationships between citizens and service organizations were mainly clientelistic: you became dependent on the professionals, and if you were dissatisfied, you could shut up or complain to a supervisor.

We discuss the administrative and democratic sides in succession, but they cannot be separated completely since they concern the same phenomena; some of the literature therefore links both sides.

4.1. Administration: Inter-governmental Management

The fragmentation at the local level opens up new roles in public administration. A number of tasks are now implemented by semi-independent public organizations or non-profit organizations, contracted out to private firms or non-profit organizations - or handled in collaboration with other communes. This requires new skills. Metcalfe (1993: 185-189) points to the need for another understanding other than the traditional bureaucratic one of sub-ordination in a hierarchy. Rather the process should be understood as one of learning within a network where the responsibility for structural change is shared between the participants. The management role of the individual organization then is to design adaptable systems rather than blueprints for reforms. The core task is to build up interorganizational cooperation rather than make subordinates comply. Metcalfe does not provide us with operational clues, but clearly dialogue is an important element. There should be more meetings, fewer closed doors, and a movement away from the paper memo requiring staff members to perform specified duties, as well as be in favor of making time-limited agreements among members of a group to transcend formal organizational boundaries.

Kooiman (1993: 252-54) stresses that in general, what he calls 'co'- arrangements require that a problem be understood as one to be handled in a communication network with multi-dimensional interdependencies, where the traditional cause-effect scheme for action must be changed into an understanding of how parts can work for a whole. The process is characterized by some chaos and discontinuity, and the only way to learn is by deciphering information fed back for future use. There is a need for stronger understanding of situational factors; where an organizational leader would base actions on rules but accept certain exceptions. The interorganizational setting offers many "exceptions" - i.e. individual cases - that may or may not be transferred into rules.

There is no doubt that these demands run counter to the perceived view of public leadership. This is linked to the closed organization with the managers in control in a fairly traditional bureaucratic structure. Consequently, much of the literature deals with the problem of the more open organization either as a specific facet of the managerial role or as a problem linked to generic theory, i.e. not interested in situational parameters. So it provides us with few clues to actual performance problems.

The particular challenge that appears to be poorly treated in the literature relates to the problem that as interorganizational communication and negotiation increases, the traditional manager loses control over what is happening in the organization. More and more people get involved in quite tricky problematics and are involved in solving them as part of a learning process along the way. This means that the manager cannot become involved in a decisive way in many problems the way he or she could previously when all such deals were made by the top. But in most of the literature, the fact that the organization is open may be acknowledged, but nonetheless quite traditional means of control based on the closed organization are being recommended to the reader.

There are a few exceptions. Agranoff (1986: 6-12) points out that in intergovernmental relations, the constitutional-legal status of the organizations must be recognized, as must the political nature of the interrelations. But the process should be understood as one moving towards an understanding of organizations as independent and equal entities rather than one where small or weak organizations are subject to subordination. First of all, a successful process requires a joint task orientation, which may be obtained by focusing on the problem at hand rather than organizational goals, and by avoiding toothless joint task forces or the like. An important skill relates to conflict resolution, e.g. by openly exchanging information of facts and then working through the differences. A zero-sum interpretation of the situation is prohibitive for reaching a solution; ability to perceive alternatives and discuss them is a must; mutual adjustment is inevitable. White (1989:528-29) confirms a need for diversity, for information exchange, for consultation and negotiation about the rules rather than just holding other units accountable to predefined rules. Managers must be attentive to those people that are closest to the implementation problems and managers must be open to learning from them.

Such skills depart from the ones necessary in the traditional inter- organizational relations - a guardedly mutual watch between organizations. They do not often differ much from the relations between departments of organizations which could otherwise be a point of reference. Such inter-departmental relations are seen as a zero-sum game with a constant risk of losing budgetary allocations or personnel to the other department. But by focusing on the substantive problem in society to be solved in stead of narrow organizational goals, we have the closest approximation to the public interest we can probably get. This should act as a first safeguard against the zero-sum perception which tends to block the interaction.

In the case of local government, there could be particular problems to be taken into account regarding the role of councillors. The leadership is dual: the elected and the hired; and the elected ones would have additional channels of influence at their disposal, first of all the political party channel. So in so far as party colleagues are involved in the leadership of other organizations in the fragmented locality, modes of power may be used that are not really recognized by many observers. It may be safely hypothesized that if a party member feels that the development at the local level is heading in a direction contrary to the ideas of his or her party, the creativity to turn the tide may be very high and include measures that hardly qualify as elements of a learning process in the normal understanding of the word. But surely participants will learn some kind of lesson.

4.2. Democratic Involvement

The factors above mainly related to the leadership of organizations include elected politicians. But fragmentation also is a challenge to the citizens who have to learn how to get involved, and use the new opportunities in a multiplicity of organizations in the locality.

Dahl (1989) lists several layers of activity that are important in a democratic society. There must be provisions for general direct participation in selection procedures linked to political representation. Furthermore, as the society becomes complex, there must be some use of expertise in the process of governance, e.g. by a professional bureaucracy. Finally, processes of governance should involve deliberative input from the population. We are watching more and new channels open for popular participation; new fora are being created, and the question then is how the population reacts to these channels.

There has been a tendency among many observers to frown upon many channels of local influence. The critique has found substantiation in the bureaucratic literature which stresses accountability obtained by clear chains of command. Likewise, citizens should have a clear understanding of who is in charge, and preferably they should only have one place to go to in case of dissatisfaction: to the political top of a unified local government system. This would yield economies of scale, and from a perspective of responsibility it would help focus political responsibility and assure a more integrated governmental response to area wide problems (Lyons & Lowery 1989:533).

But other theorists promote the idea of democratic multiplicity which they see as an antidote to the common idea of many Schools of Public Administration. They promote the idea that the best solution to problems of coordination is to create a big, common city administration, a hierarchy. Experience shows that there are many examples that small entities can handle their mission; in some cases they do not do it alone, but solve their problems in cooperation with others. Some times it is a bigger unit with spare capacity, some times it is done in cooperation with other small units, sharing important, but expensive equipment or personnel (ACIR 1988, ACIR 1992). The conceptual general building stone is that of a public economy, taking departure in Ostrom, Tiebout & Warren (1961) (but also broadening those basic ideas considerably), and splitting up the analysis of service in decisions regarding provision and production of the service (ACIR 1987:5-14). In many ways this is precisely what is happening in the three countries we have had under scrutiny, but it should be noted that the Ostrom version requires that there is genuine democratic control of the agency. It will not do just to place an organization in the locality with leadership appointed from the center.

How can we take sides? A theoretical answer depends on the type of theory that forms the basis for analysis (Warren 1992). In standard liberal-democratic theory rooted in Madison, James Mill, Weber and Schumpeter, interests are preferences which are formed outside the political sphere, and we are into what March/Olsen (1989) call the consequential model of politics which public choice theorists have refined in mathematical terms. The political system aggregates preferences, but does not change them; the electorate decides individually whether or not the political decisions made on that basis satisfy the preferences. In the alternate version, expansive democracy (Warren 1992), rooted in Rosseau, John Stuart Mill, T H Green and John Dewey, the political sphere intrudes on factors that are privatized in liberal theory; hence democracy is influencing the individual's control over self-determination and self- development. This is more on the line of March/Olsen's (1989) model of appropriateness. This kind of democracy develops values that are intrinsic to political interaction by processes of dialogue.

Dahl's ideas about democracy lie closer to the former model although he certainly is not blind to the benefits of broader democratic participation in e.g. the workplace. The processes of the traditional liberal-democratic theory do not involve the citizen except as reactions on the material needs that he or she may wish the body politic to fulfill. The process of the expansive democracy, in contrast, involves the citizen in ways that secure a future commitment to the basic values of democracy. Democracy becomes an end rather than a means. The Ostrom model may have started out in the liberal-democratic version, but as it has evolved, the present version comes somewhere in between; on the one hand it is based on a methodological individualistic basis, stressing the choices of individuals; on the other hand it stresses a more continuous mobilization of the individuals in political affairs, in particular by creating and amending institutions for collective choice (Ostrom et al. 1994). This version is far from the individual solely pursuing his or her personal goals, in isolation from other individuals, and solely acting under the constraints of an institution. The individual in the present way of thinking is actively involved in using and possibly changing institutions.

One should be careful to imply that a particular form of democracy overrules any other need in the society. So I do not want to put forth a claim that new forms of participation are on all counts more important than the contents which result from the services governed by the new democratic channel. Those two sides of the coin need to be balanced, meaning that none of them can be ignored. For instance, a systematic and continuing neglect of the needs of minorities in the provision of services would call for revision of the decision-making pattern, at least insofar as they violate human rights.

I point to two directions the development may take. The individualist version seems to be the one that most observers - and particularly Scandinavians - are afraid will be the result. The question is whether the citizen - the citoyen - will be replaced by a user maximizing nothing but short-term interests? Will the non-user refrain from involvement in politics since most matters are decided by user groups? Will local politics accordingly be reserved for the very few? It seems that most of those observers are hoping for a more community oriented or cohesive version. Here the question is whether user groups can manage to invigorate local politics, user boards serving as first experiences in politics and a number of individuals slowly getting trained for higher political office? Will local politics accordingly get renewed, less on a political party basis and more on the basis of the skilled and experienced local representative?

Note that the distinction is not the same as the liberal/expansive division above. It is the orientation of the individual that counts. If it does not go beyond own preferences and needs and if the institutional structure rewards such behavior, the locality will end up as a struggle between individual interests getting organized ad hoc with little hope for reconciliation. In that sense it is individualistic. In the cohesive version there is no neglect of individual interests, but there also is a sense of community needs and public interest which will persist even though some local dissatisfied people opt out by moving to another locality. In order for a fragmented system to better operate, I suggest that there must be a minimum of such sentiments.

The final section will expand on this precarious theme and how to analyze it within the ideas of institutional theory.

5. RESEARCH IMPLICATIONS

We can see trends towards fragmentation in the public and private sectors in many Western countries. This development in itself may challenge the very notion of public versus private, it calls for a more sophisticated understanding of systems of governance in their broadest sense - government transcending boundaries of formal organizations, and with a focus on continuing processes rather than confrontations and reconciliations of political stances. This book deals with the public side, taking departure from the formal ways of organization and describing how their boundaries have been opened up for new channels of influence. This calls for renewed interest in basic approaches.

Crook, Pakulski and Waters (1992:97-194) see the Postmodern development of the public sector at the local level as one towards the minimalist state after a process of decentralization to semi-autonomous agencies, social movements, privatization and deregulation. Although we have seen trends towards such changes above, it would be wrong to characterize the Scandinavian development as one towards a minimalist state. It would be more accurate to say that the whole concept of the state as a coherent and hierarchical construct is being challenged. It is not clear what can replace the concept except for the fact that there is some kind of network links that go back to a democratic legitimation and a public budget.

In the Postmodern society and the new public sector the individual has come much more into focus; both as a user and as a governor of services. The channels of influence have been fragmented, the consolidated local government has been sub-divided by functions and/or by local areas, and there is potential for differentiation based on the needs the users have or feel they have. The centralized fora for policy are weakened, and the result should be greater flexibility. The new forms of politics are based on the locality, and we see trends towards more functional encapsulation for a number of activities. The use of non-profit organizations as well as for- profit organizations in service delivery makes the traditional division of public and private dubious.

The attempts to explain the development are scarce. There is the Postmodern literature describing certain features mainly in the production and family spheres, but as yet little regarding the public sector. There is no doubt that the public sector faces challenges, but less agreement as to how to conceptualize those challenges. Some of them are due to the problems of the modern, industrialized society with mass unemployment, standardization of processes etc. Some are next to unknown or maybe wrongly perceived and therefore challenging in new ways; they stem from the Postmodern society in development.

The concepts of the modern social sciences - linked to the modern society - are, I will contend, of limited use. They conceptualize precisely the categories of the modern society where people are grouped according to sex, age, occupation, education, organizational and political party affiliation, religion etc. We link patterns of behavior to such categories; but in many instances we find that the links are dubious, and increasingly so. People do not vote as they used to do[2]; the sexes change behavior across the traditional gender roles; members of some congregations are openly rebellious to the religious leadership in matters of e.g. contraception; and so forth. All these categories were important in building up modern sociology and political science dealing with urbanization, political parties etc. because they mirrored the important aspects of the industrialized societies of the 19th and 20th centuries. But they are being transcended, changed or thwarted in the Postmodern society and therefore they are creating confusion rather than clarification if they are applied to the new trends.

Above I have pointed to the break-down of traditional organizational features and increased inter-organizational activity, as well as the new roles of democratic participation challenging the citizens. But we know far too little of what makes people of the 1990s involve themselves in problems of collective action, and how problems of the open organizations are handled. The traditional approaches of organizations and political parties are hardly adequate. The attempts of the various versions of "new institutionalism" (March & Olsen 1989, Ostrom et al 1994) do, despite their dif-

[2] Even if the link is there, the determining variable may be dubious. Thus at the American national elections in November 1994: Two days after the election *The New York TImes* reported that Republicans overwhelmingly voted republican, and Democrats democratic. But relatively more Democrats stayed at home, and the independents - free from party institutional constraints - voted republican. Result: landslide victory for the Republicans. The analytical problem is that fewer and fewer people are party members and hence do not relate to one of the salient features of modernity and modern political science.

ferences, help us grasp new systems of interaction across formal barriers. Let us briefly take a closer look at some basic differences between the two versions.

March/Olsen contend that there are two major alternatives in analyzing politics. The first one is based on exchanging resources - the market metaphor is clearly behind this notion. Individuals behaving rationally act to maximize their preferences in a political community that is seen atomistic and primarily serving individuals to fulfill their ends. The second one is based on a socially constructed actor who takes institutions for given and accordingly acts within the rules and practices prescribed by that institution (March & Olsen 1994:3-5).

But this way of presenting institutional research is not really helpful as an accurate description for the state-of-the-art in institutional analysis. It is as simplified as is the contrasting of actor and structure approaches like this: the aggressive market actor refraining from no legal means to get what he wants, relies on the ideas and principles of the invisible hand to take care of social problems that may come from his selfish action; versus the structural dope that has no personal intelligence but without much ado acts the way it is usually done here. None of these alternatives are attractive - neither from an empirical point of view (one can show that individual action can be of importance) nor from a theoretical point of view (how does change come about if actors always adapt?). Stronger sophistication in presenting other approaches is required.

Another problem is that March and Olsen tend to reify the institution which in my perception is to be considered a concept:

> ... in the struggle to survive, institutions transform themselves .... Surviving institutions seem to stabilize their norms, rules, and meanings so that procedures and forms adopted at birth have surprising durability (March/Olsen 1994:18)

On the other hand,

> Institutions change as individuals learn the culture (or fail to), forget (parts of) it, revolt against it, modify or reinterpret it (March/Olsen 1994:19)

So what are we to believe? Given the observations in this book on changes towards post-modern social conditions, the institutional context increasingly may not be taken for granted the way it is in the modernity. Therefore, values are less stable, institutional cultures more under change, and norms less stable as behavior shapers. So there is reason to develop theoretical approaches attuned to individual action linked to institutions rather than behavior determined by institutions.

There are attempts towards reconciling the exchange version with the institutionalist version (to use March/Olsen's concepts); in an attempt to keep the notion of people seeking to act to their own benefit, realizing that they act within institutional constraints. For lack of better concepts, this is what I call actor-in-institution, found in the more traditional type of public choice and game analysis. Another version is one of individuals pursuing some kind of goals, but realizing that there is a potential in working with the constraints and resources that are institutionally determined. Again for lack of better concepts, this is what I call actor-cum-institution, found in the recent new institutionalism based on individuals as actors.

An actor-in-institution realizes the limits of possible action and does his best to exploit the course of possible action within those constraints. This is the course taken by many "economistic" analysts, a special line of research (e.g. game theory) goes into analyzing how different institutional settings might shape the behavior of a rational actor. But one could also say that March/Olsen pursue such a line:

> Insofar as political actors act by making choices, they act within the definitions of alternatives, consequences, preferences (interests), and strategic options that are strongly affected by the institutional context in which they find themselves. (March/Olsen 1994:8).

Of course, the first eight words indicate that the political actor may not make any choice at all; still the characterization actor-in-institution seems appropriate. The actor validates action under the rules of the institution, not unlike a process of interpreting the law (March/Olsen 1994:10).

An actor-cum-institution is a socialized individual that realizes the limitations of institutional rules, but at the same time may act - if dissatisfied with the options - within those constraints to change the content or standard interpretation of such rules to allow for new courses of action. This could be done in a process of interaction with other actors where the mutual influence process is crucial for an understanding of how the outcome in terms of institutional rules-in-use (Ostrom 1990) is determined.

The difference to March/Olsen comes out in the action goals. March and Olsen advance the idea that by and large, people act in order to fulfill the ideas and principles of the institution:

> In the institutional story, people act, think, feel and organize themselves on the basis of exemplary or authoritative ... rules derived from socially constructed identities, belongings and roles. Institutions organize hopes, dreams, and fears, as well as purposeful actions. (March/Olsen 1994:5).

While there is no reason to doubt that institutions play a significant role in peoples' lives, there is reason to believe that people do act within an institution with ideas of getting some purposes, which they at least think are their own, fulfilled. Simon discusses this (with a purpose to crush Becker's use of economic rationality in all social matters):

> Everyone agrees that people have reasons for what they do. They have motivations, and they use reason (well or badly to respond to these motivations and reach their goals. (Simon 1987:25).

It is then an analytical question how to conceptualize this; Simon's purpose was to do away with the substantial rationality of neoclassical economics in favor of a more procedural rationality. The disagreement between the more individualistic and the more structural versions of the new institutionalism is not on the adaption of the substantial rationality - as indicated above, one of the most outstanding representatives of the individualistic version uses bounded rationality (Ostrom 1990). So we need to go into the formation of preferences if we are to solve - or rather shed light on - the disagreement.

My conclusion at this point is that the proper way to go is to develop the individual-cum-institution understanding; this avoids the lonely maximizer on the market that is not there, and defies the structural dope that has no capacity for change. So we need research tools that trace the actions of individuals performing a multiplicity of roles (Porter 1990) and help us organize those actions in an analytical way. They will prevent us from addressing all the traditional questions of organizational controls and motivations built on the categorizations of social sciences which are out of step with what is going on.

WE need to break down turn of the century understanding. That is, as long as the cases we have under scrutiny are dubious in those modern terms. If they are clearly modern, modern tools apply. But my feeling is that this is less and less so.

References

ACIR. 1987. *The Organization of Local Public Economies*. Commission Report A-109. Washington, D.C.: Advisory Commission on Intergovernmental Relations.

------. 1988. *Metropolitan Organization: The St. Louis Case*. Commission Report M-158. Washington, D.C.: Advisory Commission on Intergovernmental Relations.

------. 1992. *Metropolitan Organization: The Allegheny County Case*. Washington, D.C.: Advisory Commission on Intergovernmental Relations.

Agranoff, Robert J. 1986. *Intergovernmental Management. Human Services Problem-solving in Six Metropolitan Areas*. Albany, NY: State University of New York Press.

Albæk, Erik. 1995. "Reforming the Nordic Welfare Communes." *International Review of Administrative Sciences* (in print).

Beckman, Björn, Peter Bogason, Stefan Sjöblom, and Nils Aarsæther. 1991. "Policy-doktriner og ansvarsmodeller i de nordiske lande." *Nor-REFO* (3):20-28.

Cloke, Paul, ed. 1992. *Policy and Change in Thatchers's Britain*. Oxford: Pergamon Press.

Cochrane, Allan. 1991. "The Changing State of Local Government: Restructuring for the 1990s." *Public Administration* 69 (Autumn):281-302.

Crook, Stephen, and et al. 1992. *Postmodernization. Changes in Advanced Society*. London: Sage.

Dahl, Robert A. 1989. *Democracy and Its Critics*. New Haven: Yale University Press.

Giddens, Anthony. 1990. *The Consequences of Modernity*. Cambridge: Polity Press.

Hood, Christopher, and Gunnar Folke Schuppert, eds. 1987. *Delivering Public Services in Western Europe: Sharing Western European Experience of Para- government Organization*. London: SAGE.

Inglehart, Ronald. 1990. *Culture Shift in Advanced Industrial Society*. Princeton University Press. Princeton, NJ.

Kooiman, Jan. 1993. "Findings, Speculations and Recommendations." In *Modern Governance: New Government-society Interactions*, ed. Jan Kooiman, 249- 62. London: SAGE Publications.

Lyons, W. E., and Davis Lowery. 1989. "Governmental Fragmentation Versus Consolidation: Five Public-choice Myths About How to Create Informed, Involved and Happy Citizens." *Public Administration Review* 49 (November/December):533-43.

March, James G., and Johan P. Olsen. 1989. *Rediscovering Institutions. The Organizational Basis of Politics*. New York: Free Press.

------. 1994. "Institutional Perspectives on Political Institutions." Paper Presented at IPSA Conference, Berlin. mimeo.

Metcalfe, Les. 1993. "Public Management: From Imitation to Innovation." In *Modern Governance: New Government-society Interactions,* ed. Jan Kooiman, 173-89. London: SAGE Publications.

Offe, Claus. 1986. *Disorganized Capitalism*. Cambridge: Polity.

Ostrom, Elinor. 1990. *Governing the Commons: The Evolution of Institutions for Collective Action*. Cambridge: Cambridge University Press.

Ostrom, Elinor, Roy Gardener, and James Walker. 1994. *Rules, Games & Common-pool Resources*. Ann Arbor: University of Michigan Press.

Ostrom, Vincent, Charles Tiebout, and Robert Warren. 1961. "The Organization of Government in Metropolitan Areas: A Theoretical Inquiry." *American Politican Science Review* 55:832-42.

Porter, David O. 1990. "Structural Pose as an Approach for Implementing Complex Programs." In *Strategies for Managing Intergovernmental Policies and Networks,* Eds Robert W. Gage and Myrna P. Mandell, 3-28. New York: Praeger.

Rose, Lawrence E. 1990. "Nordic Free-Commune Experiments: Increased Local Autonomy or Continued Central Control." In Desmond S King & Jon Pierre (eds): *Challenges to Local Government*. London: Sage 1990, 212-41.

Simon, Herbert A. 1987 <1986>. "Rationality in Psychology and Economics." In *Rational Choice. The Contrast Between Economics and Psychology,* ed. Robin M. Hogarth and Melvin W. Reder, 25-40. Chicago: University of Chicago Press.

Smith, Steven Rathgeb, and Michael Lipsky. 1993. *Nonprofits for Hire: The Welfare State in the Age of Contracting Out*. Cambridge, MA: Harvard University Press.

Stewart, John, and Gerry Stoker, eds. 1989. *The Future of Local Government*. London: MacMillan.

Stoker, Gerry. 1991. "Introduction." In *Local Government in Europe. Trends and Developments,* Eds Richard Batley and Gerry Stoker, 1-20. London: MacMillan.

Warren, Mark. 1992. "Democratic Theory and Self-transformation." *American Political Science Review* 86 (March):8-23.

White, Louise G. 1989. "Public Management in a Pluralistic Arena." *Public Administration Review* 49 (November/December):522-32.

[illegible] 1982. [illegible] Press.

[illegible]

[illegible] 1986. [illegible]

[illegible]

[illegible] Vincent [illegible] and Robert Warren. [illegible]

[illegible]

[illegible] Press.

[illegible]

Simon, Herbert [illegible] In The Behavioral Foundations of Economic Theory, ed. Robin M. Hogarth and Melvin W. Reder. [illegible] Chicago: University of Chicago Press.

[illegible]

[illegible] The Future of Local Government. London: Macmillan.

[illegible] Local Government in Europe: [illegible] Richard Batley and Gerry Stoker. [illegible] London: Macmillan.

[illegible] 1992. "Democratic [illegible] Self-Government [illegible]

[illegible] 2: 122[illegible]

INDEX

A & B

Aalborg study, 52
accountability, 173, 181
bureaucracy, 12, 40, 96, 106, 139, 140, 157, 158, 160, 181

C

centre, 16, 20, 22, 27, 30, 32, 46, 99, 124, 126, 131, 137, 138, 147, 153
citizenship, 37, 41, 53, 54, 58, 65
civil society, 10, , 37, 39, 40, 41, 45, 52, 58, 144, 148, 149, 159, 161, 163, 166, 178
common identity, 32, 59
communes, 1, 2, 3, 4, 5, 6, 7, 9, 73-75, 170, 178
community councils, 11, 116, 119, 124, 125, 127, 140
community, , 2, 3, 6, 7, 8, 9, 10, 11, 20, 28, 31, 32, 37, 38, 42, 43, 50, 52, 53, 54, 55, 56, 57, 58, 62, 64, 65, 116, 117, 119, 120, 121, 122, 123, 124, 125, 126, 127, 128, 129, 130, 131, 132, 133, 134, 135, 137, 138, 140, 141, 142, 171, 183, 185
culture, 2, 5, 41, 53, 65, 121, 127, 129, 135, 153, 174, 186

D

decentralization, 10, 116, 119, 125, 131, 135, 136, 148, 184
decisions, 10, 11, 29, 36, 48, 54, 55, 58, 59, 60, 61, 64, 95, 96, 98, 99, 100, 111, 119, 121, 129, 133, 139, 141, 158, 164, 173, 177, 181, 182
deindustrialisation, 109
Denmark, 1, 2, 3, 5, 6, 7, 11, 44, 46, 81-82, 117, 119, 124, 129, 131, 135, 143, 144, 149, 150, 155, 162, 163, 167
direct democracy, 56, 97, 117, 118, 140, 141

E

education, 2, 4, 17, 23, 24, 64, 94, 96, 98, 147, 163, 165, 166, 175, 184

efficiency, 7, 8, 16, 20, 22, 30, 97, 98, 99, 100, 111, 173
Ejby, 117, 131, 132, 133, 134, 135, 136, 137, 141
elderly, 3, 7, 11, 12, 125, 144, 150, 151, 152, 153, 161, 162, 165, 166, 176
elites, 10, 37, 43, 44, 54, 55, 58, 59, 65

F

financial management, 32, 75
fixed capital, 17, 23
freedom of choice, 81-82
functional responsibility, 22

H

health, 2, 3, 17, 24, 40, 45, 98, 153
Herlev, 117, 129, 130, 131, 134, 135, 136, 137, 141
hospitals, 3, 17, 21
housing, 2, 24, 125

J & K

Job mobility, 175
justice, 97, 98, 99, 101
Klosterhaven, 153, 155, 156, 157, 158, 159, 160, 161, 162, 166
knowledge, 44, 47, 48, 50, 52, 54, 56, 58, 60, 61, 62, 63, 65, 125, 146, 162, 164, 167

L

lay-citizens, 37, 43, 65
leadership, 37, 38, 41, 45, 48, 50, 51, 52, 53, 54, 65, 123, 178, 180, 181, 185
life quality, 47
local democracy, 11, 28, 29, 118, 123, 132, 139, 140, 141, 177
local governance, 45, 47, 172
local opinion, 11, 117, 122, 123
locality, 9, 10, 12, 27, 38, 41, 42, 44, 47, 52, 138, 170, 171, 173, 175, 176, 177, 178, 180, 181, 183, 184

M

managerial efficiency, 8
mobilization, 120, 122, 125, 134, 135, 137, 138, 141, 182
modernization, 11, 15, 19, 21, 22, 27, 32, 54, 164
municipal council, 6, 25, 99, 119, 120, 121, 122, 123, 125, 129, 131, 133, 141
municipality, 1, 10, 20, 25, 104, 119, 121, 123, 124, 126, 128, 129, 131, 133, 135, 136, 137, 139, 141, 153, 161, 171

N & O

neighborhood councils, 1, 5, 6, 7, 10, 11, 76-78, 83-86, 172, 173, 175
Neighbourhood, 15, 98, 99
Norway, 1, 2, 3, 5, 6, 7, 11, 101, 117, 119, 124, 126, 128, 135, 136
Örebro, 79-85
organisational changes, 93, 101, 104, 111

P

personal freedom, 176
party politics, 89
policy, 4, 5, 8, 11, 20, 22, 38, 39, 40, 42, 43, 44, 45, 52, 53, 55, 56, 57, 58, 59, 64, 65, 102, 120, 129, 139, 146, 171, 172, 184
political identity, 46
political parties, 5, 10, 100, 102, 112, 119, 122, 123, 125, 126, 129, 130, 173, 185
popular participation, 16, 25, 28, 31, 171, 181
popular sovereignty, 39
privatization, 10, 32, 46, 184
public consumption, 2
public opinion, 122
public policy, 172
public sector, , 45, 94, 95, 96, 97, 99, 147, 163, 164, 165, 170, 172, 176, 177, 184
public services, 4, 8, 10, 12, 30, 95

R

refuse collection, 24
reorganization, 9, 16, 17, 19, 20, 22, 25, 31, 15
representativeness, 10, 94, 102, 104
retirement, 150
rural community, 3, 138

S

scale, 7, 9, 16, 17, 20, 21, 22, 23, 24, 25, 27, 28, 29, 30, 31, 32, 97, 99, 118, 119, 126, 170, 173, 181
schools, 2, 3, 5, 7, 11, 12, 17, 21, 95, 96, 125, 131, 144, 163, 164, 165, 177
service production, 6, 11, 97
service responsibilities, 21
social democrats, 39, 46, 79-85
social expenditures, 2
social rights, 38, 41, 53
social security, 2
state agencies, 4, 7, 94
subelites, 37, 43, 55, 58, 65
suburbanization, 15, 17, 20
Sweden, 1, 2, 3, 5, 6, 8, 10, 71-92, 101, 112, 136, 138

T, U & V

taxes, 3, 31, 45, 139
Tromsø, 117, 124, 126, 134, 135, 136, 137, 140, 141
user democracy, 78-79
voting, 30, 31, 45

W

welfare state, 2, 11, 19, 37, 38, 39, 40, 41, 42, 43, 44, 45, 46, 52, 53, 54, 65, 73-75, 98, 144, 147, 151, 172, 176
women, 101, 107, 108, 109, 110, 111, 150
working class, 109, 110